RICHARD WAGNER
AND THE NIBELUNGS

ieweile hatte Hagen
Den Schatz viel gar genommen,
Eh' der reiche König
Wieder war gekommen;
Er senkte ihn zu Loche
Allen in den Rhein.
Er wähnte sein zu genießen;
Das sollt' ihm nicht beschieden seyn.

Die Fürsten kamen wieder,
Mit ihnen viele Mannen;
Kriemhild mit Frauen und Mägden
Zu klagen da begannen
Ihren großen Schaden;
Ihnen war bitter leid.
Gerne wär' Giselher
Zu allen Treuen ihr bereit.

Da sprachen sie in Gleichem:
„Er hat viel übel gethan."
Er entwich der Fürsten Zorne,
Bis wieder er gewann

'The hoard is received into the Rhine'; illustration by Julius
Schnorr von Carolsfeld and Eugen Neureuther for Pfizer's
edition of the *Nibelungenlied*, p. 211.

Richard Wagner
and the Nibelungs

Elizabeth Magee

CLARENDON PRESS · OXFORD
1990

Oxford University Press, Walton Street, Oxford OX2 6DP

Oxford New York Toronto
Delhi Bombay Calcutta Madras Karachi
Petaling Jaya Singapore Hong Kong Tokyo
Nairobi Dar es Salaam Cape Town
Melbourne Auckland
and associated companies in
Berlin Ibadan

Oxford is a trade mark of Oxford University Press

Published in the United States
by Oxford University Press, New York

British Library Cataloguing in Publication Data
Magee, Elizabeth
Richard Wagner and the Nibelungs.
1. Opera in German. Wagner, Richard, 1813–1883
I. Title
782.1
ISBN 0–19–816190–5

Library of Congress Cataloging in Publication Data
Magee, Elizabeth
Richard Wagner and the Nibelungs/Elizabeth Magee.
Includes bibliographical references and index.
1. Wagner, Richard, 1813–1883. Ring des Nibelungen.   I. Title.
ML410.W15M3   1991
782.1'026'8—dc20   90–7512
ISBN 0–19–816190–5

Typeset by Wyvern Typesetting Ltd, Bristol

Printed in Great Britain by
Bookcraft (Bath) Ltd.,
Midsomer Norton, Avon

ACKNOWLEDGEMENTS

My special thanks are due to the staff of the Sächsische Landesbibliothek at Dresden, without whose assistance this book could not have been written. My grateful thanks go also to the Klett–Cotta Verlag; to F. A. Brockhaus *GmbH*; to the Richard–Wagner–Museum at Bayreuth; and to the Deutsches Literaturarchiv.

I am particularly indebted to Professor A. Stanforth of Heriot–Watt University and Mr P. Thornton of Newcastle University for supervising this work as a Ph.D. thesis.

My debts generally to friends, family and fellow Wagnerians for help, support, and encouragement are too numerous to mention and too great to ignore.

CONTENTS

ABBREVIATIONS

DB	C. von Westernhagen, *Richard Wagners Dresdener Bibliothek 1842–49* (Wiesbaden, 1966) (including catalogue of Wagner's library; numbers refer to numbers in this catalogue)
Dokumente	*Dokumente zur Entstehungsgeschichte des Bühnenfestspiels 'Der Ring des Nibelungen'*, ed. W. Breig and H. Fladt (Richard Wagner: *Sämtliche Werke*, ed. C. Dahlhaus, vol. xxix/1; Mainz, 1976)
HS	*Das Lied vom Hürnen Seyfrid*
ML	R. Wagner, *Mein Leben*, ed. M. Gregor-Dellin (Munich, 1976)
NL	*Das Nibelungenlied*
PE	*Poetic Edda*
Schr.	R. Wagner, *Sämtliche Schriften und Dichtungen*, 6th edn. (Volksausgabe), 12 vols. (Leipzig, n.d.)
Skizzen	*Richard Wagner: Skizzen und Entwürfe zur Ringdichtung*, ed. O. Strobel (Munich, 1930)
SnE	*Snorra Edda* or *Prose Edda*
Thss	*Thidreks saga*
Vs	*Völsunga saga*

EDITIONS AND NAMES

See Bibliography for full details of editions referred to. References to J. Grimm's *Mythologie* are to the second edition (1844); those to Fouqué's *Sigurd* are to the 1841 *Ausgewählte Werke* edition. Edda references, unless otherwise stated, are to Simrock's 1851 edition, which is the most complete. For the titles of Eddic poems, where several divergent spellings occur, Simrock's usage has wherever possible been adopted for consistency. In the cases of the chief god and main hero, both versions of their names (Wotan – Odin; Siegfried – Sigurd) are retained as appropriate.

I
Introduction

THE *Ring* text belongs to the most turbulent period of Wagner's varied life. Between the first sketch in 1848 and the completion of the poem for publication at the end of 1852 lay revolution and upheaval, flight from Dresden, and exile in Zurich, and not surprisingly the preoccupations of these years left their mark on the work in progress. *Der Ring des Nibelungen* resounds with social, political, and economic messages and the author's shifting philosophical convictions, all bearing witness to the revolutionaries, socio-economists, and philosophers whose work Wagner read or with whom he associated. Of all the contemporary movements, however, none exercised a more fundamental influence on Wagner's drama than the revival of the Nibelung legend, whose origins can be traced back to the Franks and Burgundians of the Dark Ages and which aroused such interest among Wagner's contemporaries.

For the Germany of Wagner's day 'the Nibelungs' meant first and foremost the great Middle High German epic poem, the *Nibelungenlied*. Indeed, the two were virtually synonymous, for the *Nibelungenlied* might be familiarly and affectionately referred to simply as the *Nibelungen*.[1] Written around 1200, probably in Austria, by an unknown poet and surviving in several different manuscripts, the *Nibelungenlied* clothes the more primitive saga events in a rich and detailed tapestry of the courtly society of the poet's own day.

---

[1] F. H. von der Hagen refers to 'unser grösstes Heldengedicht, die Nibelungen' ('our greatest heroic poem, the *Nibelungen*': *Die Nibelungen, ihre Bedeutung für die Gegenwart und für immer* (Breslau, 1819), 22); Wagner himself is probably using the term in this restricted sense when he describes a Leipzig student acquaintance as 'ein wahrer Haudegen aus den Nibelungen' ('a real fire-eater out of the *Nibelungen*': (R. Wagner, *Mein Leben*, ed. M. Gregor-Dellin (Munich, 1976) (*ML*), 52–3).

Although the outstanding work of German Nibelung literature, the *Nibelungenlied* was neither the final nor the definitive version of the saga. For centuries afterwards Siegfried and the Nibelungs were to crop up in all the major longer literary genres in an astonishing variety of guises, now comic, now heroic, incorporating variously elements of myth, fairy-tale, romance, and farce. Inevitably, such a medley of Nibelung literature was not achieved without considerable discrepancies and even outright incompatibilities in the version of the saga they record.

Within Germany itself the Nibelungs survived in two early products of the printing-press. *Das Lied vom Hürnen Seyfrid* has come down to us in early-sixteenth-century texts, while in 1557 Wagner's revered Mastersinger, Hans Sachs, wrote his *Hürnen Seufrid*, a tragedy in seven acts.

At some stage the Nibelung saga travelled to Scandinavia. It features in several of the poems of diverse date and style which makes up the *Poetic Edda*, surviving in manuscripts from around 1270 onwards. Snorri Sturluson included it in his thirteenth-century *Snorra Edda* (or *Prose Edda*). A similar but fuller version appeared in the *Völsunga saga*, also from the thirteenth century, while a rather different one was given in the roughly contemporaneous *Thidreks saga*.[2]

Probably awareness of the Nibelung saga never totally died out in Germany. As late as 1726 the saga reappeared in the form of a chap-book, the *Volksbuch vom gehörnten Sigfrid*, while if the Grimm brothers are to be believed Siegfried and his exploits lived on, in debased but recognizable form, in oral tradition.[3] Nevertheless, the second half of the eighteenth century saw a distinct revival of interest in the Nibelungs. In 1755 a manuscript of the *Nibelungenlied* was discovered and published, first in part,[4] then complete.[5] Academic and literary circles were intrigued, regarding the

---

[2] Other works, such as the *Rosengarten*, feature some of the same characters but hardly the same saga.

[3] J. L. C. and W. C. Grimm, *Kinder- und Haus Märchen*, 2nd. edn., 3 vols. (Berlin, 1819–22), iii. 162–77.

[4] J. J. Bodmer, ed., *Chriemhilden Rache und die Klage* (Zurich, 1757).

[5] C. H. Myller, ed., *Der Nibelungen Liet* (Berlin, 1782).

poem more or less favourably as the German equivalent of the Greek Homeric literature.[6]

The onset of Romanticism with its idealizing vision of the Middle Ages, coupled with events in the political arena of the early nineteenth century, brought the Nibelungs to the attention of a wider reading public, and for a time at the start of the century the Nibelungs enjoyed a vogue of popularity. Germany was smarting under French domination following Napoleon's military successes. The contrast between the impotence of their own divided land and the power of their mighty neighbour was inescapable to the educated German public. The movement for German unity was born, and for a while, as the tide turned in the Napoleonic Wars and hopes for liberation grew stronger, it was condoned by the authorities of Germany's particularist states. Deprived of the last vestiges of real political cohesion with the dissolution of the Holy Roman Empire in 1806, Germany looked to its cultural heritage as a rallying-point. In its quest for unity it turned to the literary works of the Middle Ages as monuments of a shared, seemingly more vigorous and glorious past, the products of an age when the empire had been a force to reckon with.

The result for the *Nibelungenlied*, the most truly ethnic among medieval German epic poems, was elevation to the role of unofficial national epic, a status it subsequently never really lost. During the first twenty years or so of the nineteenth century, as fresh discoveries of *Nibelungenlied* manuscripts were made, editions proliferated. One editor alone, F. H. von der Hagen, brought out several.[7] The poem was turned into modern German for the general readership.[8] Some of the new editions were splendidly illustrated. A special edition appeared for the freedom fighters at the front.[9] Nibelung prints adorned homes. For a while the *Nibelungenlied* featured on the curriculum of certain schools,

---

[6] A fuller account of the reception of the *Nibelungenlied* (*NL*) in the late 18th cent. and early 19th cent. can be found in Mary Thorp, *The Study of the Nibelungenlied*, (Oxford Studies in Modern Languages and Literature; Oxford, 1940).

[7] See Thorp, *Study*, p. 181 for a bibliog. of *NL* edns. during this period.

[8] Again, Thorp, *Study*, p. 185 supplies a bibliog.

[9] Zeune's *Feld- und Zeltausgabe* of 1815: Thorp, *Study*, p. 11.

while at the universities students flocked to hear A. W. Schlegel and von der Hagen lecture on it.[10]

The Nibelung fervour of the early nineteenth century left a number of tangible legacies. Apart from keeping the printing-presses busy turning out the *Nibelungenlied* in every shape and guise, the vast publishing enterprise of those years spread to other works of medieval literature. Scholars devoted themselves to tracking down and editing manuscripts, preparing new editions of early printed books, and translating into modern German from Latin, Old Norse, and indeed medieval German texts. Ambitious series of primary source editions with resonant titles such as 'Deutsche Gedichte des Mittelalters' and 'Fundgruben des alten Nordens' were announced, though like the periodical literature of the day the projects tended to be short-lived, and some ceased publication after one or two volumes.

In the course of all this activity the other major primary sources telling the Nibelung saga began to become available to the German reading public, many for the first time. In 1812 C. F. Rühs's translation of the *Snorra Edda* appeared,[11] to be followed in 1818 by that of Friedrich Majer.[12] Friedrich Heinrich von der Hagen brought out translations of the *Thidreks saga* and *Völsunga saga* in 1814–15.[13] *Das Lied vom Hürnen Seyfrid* had already appeared in a modern German version in 1811,[14] and in 1825 an original-language edition was published.[15] The 1840s saw a reissue of the *Volksbuch*.[16]

The *Poetic Edda*, linguistically the most difficult text, was rather longer in appearing. Translations of individual poems, sometimes only a few trial verses, had been making their appearance in collections and periodical literature since the close of the previous century.[17] The first separate volumes to

---

[10] Further details in Thorp, *Study*, pp. 7–11.

[11] *Die Edda* (Berlin, 1812).

[12] *Mythologische Dichtungen und Lieder der Skandinavier* (Leipzig, 1818).

[13] *Nordische Heldenromane*, vols. i–iv (Breslau, 1814–15). The final (5th) vol. did not appear until 1828.

[14] F. H. von der Hagen, ed., *Der Helden Buch*, vol. i (Berlin, 1811).

[15] F. H. von der Hagen and A. Primisser, eds., *Der Helden Buch in der Ursprache* (Deutsche Gedichte des Mittelalters, vol. ii/2; Berlin, 1825).

[16] K. Simrock, ed., *Die deutschen Volksbücher*, 13 vols. (Frankfurt-on-Main, 1845–67); vol. iii, *Der gehörnte Siegfried* (1845).

[17] e.g. F. D. Gräter's trans. in *Nordische Blumen* (Leipzig, 1798).

come out were the almost simultaneous translations of Eddic heroic poems, the 'Heldenlieder', by von der Hagen (1814)[18] and the brothers Grimm (1815).[19] Volumes of selected mythological poems—the 'Götterlieder'—appeared at intervals over the ensuing years, but not until Karl Simrock published his *Edda* in 1851 was the complete *Poetic Edda* available in German translation.

Hard on the heels of their publishing activity came the Nibelung scholars' second major contribution, in the shape of commentary and criticism. Perhaps nothing so characterizes the era as the Nibelung studies it produced. This was a field where Romantic vision and scientific research entered the lists, with victory going now to the one, now to the other. Hermann Schneider in his fascinating essay 'Richard Wagner und das germanische Altertum' sums up the academics of the day into two classes, the 'great scholars' and the 'muddleheads' or, more picturesquely, the 'good and bad fairies' of Germanic philology,[20] whereby Karl Lachmann and the Grimm brothers usually qualified for the first category. Among the bad fairies were Carl Göttling, von der Hagen in his less restrained moments, and, by common consent of contemporaries and posterity alike, F. J. Mone.[21]

The result was an intriguing mixture of serious scholarship and erudite fantasy. Much solid groundwork was accomplished during these years on questions such as the authorship and dating of the *Nibelungenlied*, manuscript traditions, textual variants, and the prehistory of the poem. Karl Lachmann brought out his controversial 'Lieder-theory'. The great debate was opened on the historical versus mythological origins of the saga.

With the recovery of further Nibelung texts, scholars faced the challenge of the variations and contradictions in the saga that they presented. Romanticism was in its heyday in seats of learning as well as in literary salons, and the belief in the

[18] *Die Edda Lieder von den Nibelungen* (Breslau, 1814).

[19] *Lieder der alten Edda* (Berlin, 1815).

[20] 'grosse Forscher' and 'Wirrköpfe'; 'gute und böse Feen' (*Kleinere Schriften zur germanischen Heldensage und Literature des Mittelalters* (Berlin, 1962), 109).

[21] Mary Thorp writes that Mone 'seems to have been generally despised by the leading critics of the period' and notes particularly the opinion of Lachmann and the Grimm brothers (*Study*, p. 32).

intrinsic Oneness of the saga triumphed over the obstacles of the material. It was an age in which inner necessity was widely recognized as the all-compelling argument.[22] One common feature and a little ingenuity could suffice to find the most diverse persons and events identical. Alternatively, scholars might select what seemed to them a convincing account of events from the profusion of material available and discard the rest as interpolations, later additions, borrowings, the results of social change, and so forth.[23]

At least equal zeal was devoted to interpreting the saga, and it was here especially that Schneider's 'bad fairies' distinguished themselves. Led on by their quest for the unity of existence, scholars such as Göttling, Mone, and von der Hagen discovered history in poetry,[24] Siegfried in Christ,[25] and in the Nibelung saga the common myth of all mankind.[26]

The re-emergence of the Nibelungs in Germany evoked a response from the literary world too. Whether motivated by enthusiasm for the subject or sober calculation of its topical appeal, patriotic fervour or artistic challenge, poets and playwrights felt drawn to create their own versions of the legend. The *Nibelungenlied* itself was naturally far and away the most popular source on which they based their works, but it did not have a monopoly, and some writers attempted a synthesis of more than one. The first substantial work to appear was Friedrich Baron de la Motte Fouqué's dramatic poem based on the *Völsunga saga*, *Der Held des Nordens*, a trilogy first published in Berlin in 1810. Further dramatic works followed, including F. R. Hermann's *Die Nibelungen*

[22] See e.g. C. W. Göttling, *Nibelungen und Gibelinen* (Rudolstadt, 1816), 22: 'Der innere und also gründlichste Beweis dafür ist jene schon dargelegte Ansicht von der innerlich nothwendigen Gestaltung des Heldenthums vom Weltlichen zum Geistlichen.' ('The inner and therefore [sic] most fundamental proof for this lies in the view we have already put forward on the transformation through inner necessity of heroic life from the profane to the secular.')

[23] This was the approach adopted by Wilhelm Grimm in the section 'Ursprung und Fortbildung' of his *Die deutsche Heldensage* (Göttingen, 1829), 335–94, and by K. Lachmann in his essay of 1829, 'Kritik der Sage von den Nibelungen', repr. in his *Zu den Nibelungen und zur Klage* (Berlin, 1836), 333–49.

[24] Thus Göttling in *Nibelungen und Gibelinen*.

[25] Or Christ in Siegfried. See e.g. the section on solar myth in F. J. Mone's *Einleitung in das Nibelungen-Lied zum Schul- und Selbstgebrauch bearbeitet* (Heidelberg, 1818), 67 ff.

[26] Most particularly in F. H. von der Hagen, *Die Nibelungen*.

(Leipzig, 1819), J. W. Müller's *Chriemhilds Rache* (Heidelberg, 1822), and two from the 1830s: E. Raupach's *Der Nibelungen-Hort* (Hamburg, 1834) and C. Wurm's *Die Nibelungen* (Erlangen, 1839). The year 1854 saw the première of the first Nibelung opera, *Die Nibelungen*, with music by Wagner's acquaintance from his Leipzig and Riga days, Heinrich Dorn.[27] Drama apart, the Nibelungs inspired a certain amount of poetry, including Siegfried poems by Uhland[28] and Fouqué[29] and sections of Simrock's *Amelungenlied*.[30] In prose there was Guido Görres's *Der hürnen Siegfried und sein Kampf mit dem Drachen* (Schaffhausen, 1843).

The ripples of Nibelung interest eventually became caught up in the cross-currents of broader contemporary movements. Scholars were taking an increasing interest in the whole field of Germanic antiquity, and much research was initiated during these years into the society, language, and beliefs of their ancestors. The scene was dominated by the towering achievements of Jacob Grimm, whose *Deutsche Rechtsalterthümer* (Göttingen, 1828), *Deutsche Mythologie* (Göttingen, 1835), and on the philological side *Deutsche Grammatik* (Göttingen, 1819–37) and *Geschichte der deutschen Sprache* (Leipzig, 1848) were all pioneering works in their time.

Folklore and folk culture, which the Romantic poets had found aesthetically so inspiring, fell into the catchment area of the same scholars. The Grimm brothers were again in the forefront, with their painstaking sifting of chronicles and other, mainly written, sources, from which they gleaned their *Deutsche Sagen* (Berlin, 1816–18), and the gathering and recording of the oral tales which form their *Kinder- und Haus Märchen* (Berlin, 1812–15), known to us as 'Grimms' Fairytales'. For the Grimm brothers these tales represented no less than late variations of ancient Germanic myths; in the

---

[27] A fuller bibliog. can be found in Thorp, *Study*, pp. 189 ff.
[28] 'Siegfrieds Schwert', written in 1813, appeared in various edns. of Uhland's collected poems.
[29] 'Der gehörnte Siegfried in der Schmiede' first appeared in *Europa*, 2 (1803), 82–7.
[30] K. Simrock, *Das Heldenbuch*, vols. iv–vi (Stuttgart and Tübingen, 1843–9).

introduction to the *Märchen* they write: 'As far as the content itself is concerned . . . a basis, a significance, a nucleus can be readily recognized. Thoughts on the divine and the spiritual aspects of life are preserved here: ancient belief and doctrine steeped in the epic element . . . and given corporeal form.[31] In one particular group of fairy-tales they spotted reflections of the Nibelung saga.[32]

The Napoleonic Wars came to an end and Germany settled back into a nation of particularist states, reduced in number but still with no effective central government. As the euphoria of victory wore off reaction set in. The movement for national unity, so useful in driving out the French, was now officially frowned upon. Popular interest in the Nibelungs waned, and during the 1820s and 1830s the saga retreated to the academic domain. There, among the students, the symbolic role of the *Nibelungenlied* was kept alive; and in the years leading up to the revolutions of 1848–9 the Nibelungs once again became a rallying-call for the beleaguered movement for German unity. During the 1840s the number of *Nibelungenlied* editions, both original- and modern-language, rose dramatically. In 1843 Guido Görres invoked Siegfried's return to restore his native land to unity and strength[33] and Anton Alexander Graf von Auersperg, writing under the pseudonym Anastasius Grün, published his political satire *Nibelungen im Frack*.[34] The following year F. T. Vischer published his 'Vorschlag zu einer Oper', in which he called for a German national opera based on the *Nibelungenlied* and even provided a specimen scenario,[35] while in the appendix of 1849 to his *Heldenbuch* Karl Simrock

---

[31] 'Was den Inhalt selbst betrifft . . . es lässt sich darin ein Grund, eine Bedeutung, ein Kern gar wohl erkennen. *Es sind hier Gedanken über das Göttliche und Geistige im Leben aufbewahrt: alter Glaube und Glaubenslehre in das epische Element . . . getaucht und leiblich gestaltet.*' *Märchen*, i. p. xxvi.

[32] *Märchen*, No. 90, 'Der junge Riese' ('The Young Giant'); No. 91, 'Dat Erdmännekin' ('The Gnome'); No. 92, 'Der goldene Berg' ('The King of the Golden Mountain'); No. 93, 'Die Rabe' ('The Raven'); No. 94, 'Die kluge Bauerntochter' ('The Peasant's Wise Daughter'). In the notes to these tales (iii. 162–77) the Grimm brothers comment on the relationship in each case to the Siegfried saga.

[33] *Der hürnen Siegfried*, p. 80.

[34] Leipzig, 1843.

[35] *Kritische Gänge*, vol. ii (Tübingen, 1844), 399 ff.

was again describing the *Nibelungenlied* as 'our national epos' and 'the greatest treasure of our people'.[36]

Such was the Nibelung climate into which Wagner's *Der Ring des Nibelungen* was born. The text of the *Ring* drama really falls into three distinct writing-phases. The summer and autumn of 1848 saw Wagner intensely occupied with the Nibelungs and related themes. Some of the accumulated overload was discharged in Wagner's essay *Die Wibelungen*. Meanwhile, he sketched out his own version of the saga as *Die Nibelungen-sage (Mythus)*, which he completed on 4 October 1848,[37] and later published as *Der Nibelungen-Mythus* with the qualifying subheading 'Sketch for a Drama'.[38] For the time being, however, only the final and most fully developed section was turned into an opera libretto, *Siegfrieds Tod*, the forerunner of *Götterdämmerung*, in November of the same year.

After the work of 1848 Wagner's creative writing went through a quiescent phase, or rather a non-productive one, for during the revolution year of 1849 Wagner was feverishly occupied with dramatic projects and sketches, none of which were ever realized. Not until he had re-established himself in exile in Zurich did Wagner return to his Nibelung drama. In the spring of 1851 he was suddenly inspired to write a second Siegfried drama, a contrasting work, as he explains in his correspondence of the period[39]—a light, comic Young Siegfried opera to precede the earnest tragedy of *Siegfrieds Tod*.

It had long been Wagner's ambition to write such an opera. As far back as 1837, becalmed on the voyage to Riga, he had picked up the chap-book *Till Eulenspiegel* and first conceived

---

[36] 'unser Nationalepos'; 'den grösten Hort unseres Volkes': vi. 397.

[37] Information on the dating of Wagner's *Ring* texts is generally taken from O. Strobel, ed., *Richard Wagner: Skizzen und Entwürfe zur Ringdichtung* (Munich, 1930) (*Skizzen*). For *Die Wibelungen* the dating followed is that established by John Deathridge, Martin Geck, and Egon Voss in the *Wagner Werk-Verzeichnis* (Mainz, London, New York, and Tokyo, 1986), 329.

[38] 'Als Entwurf zu einem Drama': R. Wagner, *Sämtliche Schriften und Dichtungen*, 6th edn. (Volksausgabe), 12 vols. (Leipzig, n.d.) (*Schr.*), ii. 156–66.

[39] Letters to Theodor Uhlig, 10 May 1851 and Ferdinand von Ziegesar, 10 May 1851: *Dokumente zur Entstehungsgeschichte des Bühnenfestspiels Der Ring des Nibelungen*, ed. W. Breig and H. Fladt (Richard Wagner: *Sämtliche Werke*, ed. C. Dahlhaus, vol. xxix/1; Mainz, 1976) (*Dokumente*), 42–3.

the idea of an 'authentic German comic opera'.[40] The 'comic, well-nigh popular material for a "Young Siegfried" ' realized his ambition;[41] and as he worked on his Young Siegfried drama Wagner thought back to the Riga voyage and his original comic-opera inspiration.[42] The third phase followed soon after. By the autumn of the same year Wagner had perceived and accepted that half measures could not satisfy him. 'I now see', he wrote to Uhlig on 12 November, 'that in order to be completely comprehensible from the stage I shall have to dramatize the whole myth.'[43] A third drama, *Die Walküre*, and a prologue, *Das Rheingold*, were conceived, completing the dramatization of Wagner's original *Mythus*. They occupied him for about a year. Then, in the autumn of 1852, Wagner undertook a revision of his Young Siegfried drama and a thorough overhaul of *Siegfrieds Tod*, which now became to all intents and purposes the present *Götterdämmerung*. Apart from a few further changes[44] Wagner had now completed the text of his *Ring*, which first appeared in a limited, privately printed edition at the beginning of 1853.

The commencement of the *Ring* at a time when the Nibelungs had assumed such significance in German awareness was obviously no coincidence. Although Wagner was

---

[40] 'echt deutschen komischen Oper': *ML*, pp. 152–3.
[41] 'heitere und fast populäre Stoff zu einem "jungen Siegfried" ' (letter to Theodor Uhlig, 12 Nov. 1852, *Dokumente*, p. 58).
[42] *ML*, p. 153.
[43] 'Jetzt sehe ich, ich muss, um vollkommen von der Bühne herab verstanden zu werden, den ganzen Mythos plastisch ausführen.' (Letter to Theodor Uhlig, 12 Nov. 1851, *Dokumente*, p. 58).
[44] Notably to some sections of *Siegfried* and the ending of *Götterdämmerung*. Much has been written about how, how often, and when Wagner altered the ending of the final drama in the cycle, from which it transpires that some changes to *Siegfrieds Tod* were undertaken even before the 1852 *Ring* revision. See e.g. *Skizzen*, pp. 58–60; W. Ashton Ellis, 'Die verschiedene Fassungen von *Siegfrieds Tod*', *Die Musik*, 3 (1904), 239–51, 315–31; O. Strobel, 'Zur Entstehungsgeschichte der Götterdämmerung', *Die Musik*, 25 (1933), 336 ff; C. Dahlhaus, 'Über den Schluss der Götterdämmerung', *Richard Wagner, Werk und Wirkung* (Regensburg, 1971), 97–115; S. Spencer, 'Zieh hin! Ich kann dich nicht halten', *Wagner*, NS, 2 (1981), 98–120. Since both the post-1853 changes and the pre-1852 alterations to *Siegfrieds Tod* resulted from dramatic and philosophical considerations and not from new influence from contemporary Nibelung interest, the exact timing of them need fortunately not conern us here.

till in his infancy at the time of the first wave of Nibelung
enthusiasm, the products of his childhood years later made
their way on to his reading-list and into his *Ring*. The revival
of the 1840s, when the Nibelungs again became a watchword
in the mounting nationalist and revolutionary fervour,
reached Wagner at his most receptive. On the shelves of his
Dresden apartment were no fewer than four different editions of the *Nibelungenlied*,[45] while over his desk hung the
exquisite Nibelung print, the title-page to the *Nibelungenlied*
by Cornelius.[46] Swept along himself by the revolutionary
movement, Wagner, like Auersperg, found in the Nibelungs
a suitable vehicle for topical commentary and poured into his
Mythus and the *Wibelungen* essay which followed it his current views on monarchy and wealth.

Wagner's interest in the Nibelungs was much more than a
passing fashion, however. It was rather one aspect of a long-
standing and deep-rooted interest in the literature of the Middle Ages and the remoter Germanic past. Perhaps we can
trace the beginnings of this interest already in the folk
literature Wagner was reading on the Riga voyage of 1837.
Wagner himself dated the dawn of his pursuit of Germanic
philology to his Paris days. In his final winter there his friend
Lehrs had brought him a copy of the journal of the
Königsberger Deutsche Gesellschaft; an article on the Wart-
burgkrieg revealed to Wagner 'the German Middle Ages in a
significant colouring, of which I had previously received no
intimation'.[47] In the same issue he apparently found an article
on Lohengrin which opened up a whole new world to him.[48]

On his return to Germany, where he had found a perma-
nent post at the Dresden court, Wagner took up his new-
found interest more seriously. In his autobiography he tells
of the fruitful summer of 1843 which he spent, ostensibly
taking the cure, at Teplitz. Here he read for the first time the
Grimm brothers' *Deutsche Sagen* and above all Jacob Grimm's
difficult and problematical but endlessly inspiring *Deutsche*

---

[45] See ch. 2.
[46] *ML*, p. 274.
[47] 'das deutsche Mittelalter in einer prägnanten Farbe, von welcher ich bis dahin
eine Ahnung erhalten hatte' (*ML*, p. 224).
[48] Ibid.

Mythologie. The reward for struggling through the latter work was that Wagner found himself

> spellbound by a marvellous enchantment: the most threadbare material spoke to me with the immemorial voice of my native land, and soon all my perceptive faculties were taken up with images, forming themselves in me with ever-increasing clarity into an anti-cipation of recapturing a long-lost and constantly resought consciousness.[49]

In the autumn of the same year Wagner established a household for himself at Dresden in the style he regarded as commensurate with his new 'Kapellmeister' dignity. Among his top priorities was a library, of which he writes in *Mein Leben*: 'In it Old German literature held pride of place and the most closely related medieval literature generally . . . Along-side these ranked good historical works on the Middle Ages and on the German people in general . . .'[50] According to the catalogue Curt von Westernhagen has prepared of Wagner's Dresden library,[51] well over half the titles on Wagner's shelves come into one or other of these categories.

Thus was born a love of medieval literature and an interest in the Germanic past which were to govern all Wagner's mature work. When therefore Wagner turned to the topical Nibelung theme as the subject of his new drama it was in a spirit of intense involvement far removed from the merely imitative. His independent outlook manifests itself with characteristic sharpness in his fundamental rejection of the *Nibelungenlied*, the prized German national epic, as a source for his drama. In his 'Mitteilung' he writes of his efforts to restore the figure of Siegfried to a purer, more primitive form and adds: 'Only then did I recognize the possibility of turning him into the hero of a drama, which had never occurred to

---

[49] 'durch wunderbaren Zauber festgebannt: die dürftigste Überlieferung sprach urheimatlich zu mir, und bald war mein ganzes Empfindungswesen von Vorstel-lungen eingenommen, welche sich immer deutlicher in mir zur Ahnung des Wiedergewinnes eines längst verlorenen und stets wieder gesuchten Bewusstseins gestalteten.' (*ML*, p. 273.)

[50] 'Am vorzüglichsten war hierin die altdeutsche Literatur vertreten und das ihr zunächst verwandte Mittelalterliche überhaupt . . . Hieran reihten sich die guten Geschichtswerke des Mittelalters sowie des deutschen Volkes überhaupt . . .' (*ML*, p. 274).

[51] *Richard Wagners Dresdener Bibliothek 1842–49* (Wiesbaden, 1966). (*DB*).

me so long as I knew him only from the medieval *Nibelungenlied*.'[52]

Nor indeed was Wagner slavish in his adherence to any given source. In the autumn of 1851, when he was just embarking on *Das Rheingold* and *Die Walküre*, he wrote to his friend Uhlig asking for the *Völsunga saga* to be sent out to him so that he could glance through it again. He is quick to point out, though, that he wants the *Völsunga saga* 'not to model myself on it (you'll soon see how *my* poem relates to the saga), but in order to recall precisely everything I had previously thought out in the way of individual details.'[53]

The letter to Uhlig is very revealing of the way Wagner approached his Nibelung studies: existing source material served to inspire, not to dictate the form his own work was to take. It also gives an insight into Wagner's reading methods which is confirmed by a later remark Wagner made, quoted in the interesting chapter 'Wie er las' of Curt von Westernhagen's *Dresdener Bibliothek*. To Princess Marie Wittgenstein he wrote in 1857: 'What happens with me is that I seldom actually read what's in front of me, but rather what I read into it.'[54] Reading for Wagner was essentially a two-way process in which his receptive and imaginative powers worked in tandem and what he read became swallowed up in the creative activity it stimulated.

And Wagner did read, avidly, whatever he could lay hands on on the subject of the Nibelungs. In the 'Epilogischer Bericht' to the *Ring* he writes of the 'in-depth studies I made of the existing myth and which first revealed to me its characters in the only light of value for a drama . . .'[55] Wagner reaped full benefit from the wealth of contemporary

[52] 'Erst jetzt auch erkannte ich die Möglichkeit, ihn zum Helden eines Dramas zu machen, was mir nie eingefallen war, so lange ich ihn nur aus dem mittelalterlichen Nibelungenliede kannte.' ('Eine Mitteilung an meine Freunde', *Schr.*, iv. 312.)

[53] 'nicht um nach ihr zu bilden (Du wirst leicht finden wie sich *mein* Gedicht zu dieser Sage verhält), sondern um mich wieder genau zu erinnern, was ich an einzelnen Zügen schon einmal konzipiert hatte.' (Letter to Theodor Uhlig, 12 Nov. 1852, *Dokumente*, p. 58.)

[54] 'Mir geht es nun einmal so, dass ich selten eigentlich das lese, was vor mir steht, sondern das, was ich hineinlege.' (*DB*, p. 11.)

[55] 'eingehenden Studien, welche ich über den vorliegenden Mythus machte, und welche mir die Gestalten desselben zuerst in einem für das Drama einzig wertvollen Lichte zeigten . . .' (*Schr.*, vi. 262).

Nibelung publications, not least from the access they provided to primary source material. The efforts of the editors
and translators of his day had supplied not only his four
different editions of the *Nibelungenlied* but also, in the shape
of the Eddas, the *Völsunga saga, Thidreks saga,* and *Das Lied
vom Hürnen Seyfrid,* alternative saga sources for the dissenting
composer. He followed the current research into Germanic
antiquity, fairy-tale, folklore, and myth and read the Grimm
brothers' *Deutsche Sagen,* their *Märchen,* and their monumen-
tal reference works.

Many of the primary source editions Wagner used came
ready-equipped with their own critical material. These apart,
Wagner drew on a representative selection of the abundant
secondary literature then available. As an artist he sought
illumination rather than information among the scholars, and
their inspirational qualities were of more value to the author
of the *Ring* than considerations of academic soundness. Lachmann and Mone, good and bad fairies alike, featured on his
reading-list, and for his purposes it mattered very little into
which category his chosen authorities fell.

Wagner, always eager to acknowledge his debt to the
philologists, was more reticent on the subject of what he
owed to fellow Nibelung poets and dramatists. Only one
such work finds mention on the list Wagner left of his *Ring*
sources, and then only obliquely: the *Amelungenlied,* which
comprises volumes iv–vi of Simrock's *Heldenbuch,* contains
his reworking of the Nibelung saga. When Wagner finally
spoke out about other contemporary Nibelung dramatists in
the 'Epilogischer Bericht' it was with disdain for their
research efforts and with the suggestion that any influence
had been entirely from him to them.[56]

As for stage works prior to the appearance of his *Ring* text,
Wagner writes that he was aware of only one 'theatre piece'
based on the *Nibelungenlied,* Ernst Raupach's *Der Nibelungen-
Hort,* which had been performed in Berlin.[57] There is no need
to doubt the veracity of Wagner's statement; it does,
however, require some qualification. Fouqué's *Der Held des
Nordens,* for instance, was almost certainly known to

[56] *Schr.,* vi. 262–3.
[57] 'Theaterstück' (*Schr.,* vi. 262).

Wagner;[58] but it was based on the *Völsunga saga*, not the *Nibelungenlied*. Besides, it was intended as a dramatic poem, not as a stage work, and had indeed never been performed, and so could perhaps not properly be described as a piece for the theatre. Wagner may have been truly unaware of the existence of some of the contemporary Nibelung dramatizations, and his comment does not extend to reworkings of the saga other than those in dramatic form. We shall in fact discover that Wagner was better acquainted with the work of predecessors in the field than his account in the 'Epilogischer Bericht' would have us believe, and that, however he might rate their scholarship,[59] he derived considerable benefit from their pioneering efforts.

Wagner also responded to visual stimuli. Apart from the Cornelius print over his desk he possessed two of the illustrated *Nibelungenlied* editions then in vogue.

In Part II of the present study we shall be examining in detail the various influences exercised on the *Ring* text by the Nibelung revival of Wagner's day, phase by phase, following approximately the three major writing-periods of the poem. The one exception to this rule will be in the chapter on the Young Siegfried drama, where for the sake of convenience the later alterations to the text will be considered out of chronological order along with Wagner's original version of the text. It will also be noted that our subject is the influence of the Nibelung activity of Wagner's contemporaries, not of the primary sources—the *Völsunga saga* and Eddas, the *Nibelungenlied*, *Das Lied vom Hürnen Seyfrid*, and *Thidreks saga*—themselves. These latter have already been widely, if not fully comprehensively, dealt with by others, notably Ernst Koch,[60] Wolfgang Golther,[61] Jessie Weston,[62] and, in a

---

[58] See ch. 4.

[59] Uhland and Simrock, two of Wagner's predecessors, were at the forefront of contemporary scholarship; but perhaps Wagner chose to ignore these particular contemporaries when making his derogatory remarks.

[60] *Richard Wagner's Bühnenfestspiel Der Ring des Nibelungen in seinen Verhältnis zur alten Sage wie zur modernen Nibelungendichtung betrachtet* (Leipzig, 1876).

[61] *Die sagengeschichtlichen Grundlagen der Ringdichtung Richard Wagners* (Berlin, 1902).

[62] *The Legends of the Wagner Dramas* (London, 1896).

recent uncompleted work, Deryck Cooke.[63] We shall nevertheless sometimes be making quite legitimate reference to the
primary sources. Partly this will be to supply continuity and
make plain the context within which Wagner's contemporaries operated. It will also illustrate the important role
played by the editors and translators of the primary
literature, and to demonstrate how Wagner's Nibelung
drama was influenced by the availability or non-availability
of the source material.

The potential rewards of such a study are numerous. In the
first place, it will enable us to see the *Ring* in its intellectual
and literary context. Paradoxically, it is only in relation to the
generality that the distinctive merits of an individual work
become fully apparent. So one gain will be in uncovering the
background to Wagner's poem, against which the achievements of the *Ring* will stand out in sharper focus and clearer
perspective.

Secondly, it will afford insight into Wagner's mind, both
its contents and its mode of operation. We shall be better able
to appreciate the full extent of his preoccupation with the
Nibelungs and the intensity with which he pursued his subject. It will offer a chance to observe the creative genius of
Wagner the dramatist at work: ideas taking root, half-
remembered inspirations from the past bearing late fruit, dry
stems of scholarship breaking forth into fresh redeeming
green, and the whole quickened into new, organic life. In
short, we shall have a practical demonstration of the kind of
process Curt von Westernhagen portrays in the chapter 'Wie
er schuf' of his *Dresdener Bibliothek.*[64]

Furthermore, we shall come to a better understanding of
the *Ring* itself. No one would pretend that the text of
Wagner's Nibelung drama, written in so many stages over a
period of several years and incorporating such diverse
perspectives, is without its puzzles. Not all its difficulties will
yield to perusal in the light of contemporary Nibelung inter-

[63] *I saw the World End* (London, 1979). There are in addition numerous other
works on the subject, incl. H. Lebede, *Richard Wagners Musikdramen. Quellen, Entstehung, Aufbau*, vol. ii (Dresden, n.d.); H. von der Pfordten, *Richard Wagners
Bühnenwerke in Handlung und Dichtung nach ihren Grundlagen in Sage und Geschichte*,
7th edn. (Berlin, 1920).

[64] Pp. 18–47.

est, but a number certainly will. And by learning what Wagner himself knew on the subject, many nuances will be revealed in the *Ring* text which otherwise tend to go unperceived, ambiguities and subtleties of meaning which were all part of Wagner's creative intention.

In view of the fact that our subject promises to be such a fruitful field of study it is perhaps surprising that it has to date commanded so little attention. Apart from a few isolated references in other works, which will be dealt with as they occur, only the merest handful of scholars have addressed themselves in any way seriously to the subject.

The earliest such work to appear, Ernst Koch's *Richard Wagner's Bühnenfestspiel Der Ring des Nibelungen in seinem Verhältnis zur alten Sage wie zur modernen Nibelungendichtung betrachtet* (Leipzig, 1876), handles the aspect of other contemporary reworkings of the Nibelung saga. As such it is interesting and instructive; but since no attempt is made to trace influence in either direction, or even to distinguish between what Wagner knew and what he did not know, it can offer no insight into Wagner's mind or the genesis of the *Ring*. Koch's book is useful only in supplying something of the contemporary backcloth to Wagner's work and remains, even within its limited terms of reference, essentially two-dimensional. Much the same is true of Ernst Meinck's *Die sagenwissenschaftlichen Grundlagen der Nibelungendichtung Richard Wagners* (Berlin, 1892), in which Wagner's work is treated as one strand in the vast tapestry of studies in Germanic mythology.

More to the point is Hermann Schneider's paper 'Richard Wagner und das germanische Altertum', from which we have already quoted pithy comments. Schneider deals with the influence of contemporary scholarship in the field of Germanic philology on Wagner's dramatic output, more especially on the *Ring*, and provides a stimulating introduction to the subject. It is, however, no more than an introduction; the paper's brief scope barely suffices for the author to indicate the type of relationship that existed, and his study does also, perhaps inevitably, contain certain substantial inaccuracies.

Most recently, Deryck Cooke's *I saw the World End*

contains sections on Wagner's use of certain works of secondary literature in the *Ring*, including some quite detailed examples. Cooke also makes greater headway than his predecessors in establishing which editions of the *Edda* Wagner used. In the vast-ranging project Cooke was envisaging there was obviously no opportunity to develop the subject fully. In any case, his work remained sadly unfinished.

Probably the most useful work all round to date is Wolfgang Golther's *Die sagengeschichtlichen Grundlagen der Ringdichtung Richard Wagners* (Berlin, 1902). As the title suggests, Golther is concerned first and foremost with Wagner's use of primary source material (he unearths some interesting minor ones), but he also mentions influence from secondary literature and the contribution of other contemporary Nibelung poets. His work by no means exhausts the subject but it is on the whole remarkably accurate.

The great obstacle to a really comprehensive study of the influence of contemporary Nibelung interest on Wagner's *Ring* poem had undoubtedly been the absence until now of a reliable bibliography. No one has been quite sure what Wagner read, or just how much he knew of what was happening on the contemporary Nibelung scene. Wagner did of course leave some clues, and the one scholars have seized on most eagerly is a letter he wrote in 1856 to Franz Müller, who was contemplating writing a book on the *Ring* text. In response to Müller's request for information on his sources Wagner replied:

I do not know whether I shall ever finish my *Nibelungen* and, even making the most favourable assumptions, could not therefore give any indication when your work should come out. So if you still want to go ahead with it now, on the off chance, I will for the time being just list the sources which I studied at the time and which prepared me for my subject.

The letter was accompanied by the following list:

1. *Der Nibelunge Noth und Klage*, edited by Lachmann.
2. *Zu den Nibelungen* etc. by Lachmann.
3. Grimm's *Mythologie*.

4. Edda
5. *Volsunga-Saga*, translated by Hagen (Breslau)
6. *Wilkina- und Niflungasaga* [*Thidreks saga*], ditto.
7. *Das deutsche Heldenbuch*, old edition; also modernized by Hagen. — Adapted in 6 volumes by Simrock.
8. *Die deutsche Heldensage* by Wilhelm Grimm
9. *Untersuchungen zur deutschen Heldensage* by Mone (very important)
10. *Heimskringla*, translated by Mohnike (I think!) (not by Wachter — bad)[65]

Even this brief list raises a number of questions. What, for instance, did Wagner mean by 'Edda'? Was he referring to the *Poetic Edda*, *Snorra Edda*, or both? And if he had the *Poetic Edda* in mind, exactly which translation? It has already been pointed out that prior to the appearance of Simrock's *Edda* in 1851 no complete translation of the *Poetic Edda* was available, and each of the volumes published by Simrock's predecessors contained a different selection of poems. Or should we believe that Wagner read the *Poetic Edda* in the original Old Norse? Wagner's reference to the *Heldenbuch* likewise calls for a certain amount of clarification; as it stands, it is not clear just how many different editions he is referring to.

In any case, it soon becomes clear that the list Wagner sent to Müller is not exhaustive. Indeed, the tone of the accompanying letter suggests that the list was rather a provisional jotting than a comprehensive bibliography. Wagner was simply trying to recall, at a distance of some years and several hundred miles, the works which had made the most durable impression during the gestation of the *Ring*, and was perhaps in some cases not uninfluenced by what he had come across more recently in Zurich.

Wagner scholars have indeed always tended to assume that the Müller list needs supplementing and have come forward with their own suggestions. Most of these must class as bold guesswork: often plausible, sometimes impressive, but short on evidence. In some cases choices have been justified from a

---

[65] 'Ich weiss nicht ob ich je meine Nibelungen beenden werde, und könnte, selbst unter den günstigsten Annahmen, somit jetzt nicht den Zeitpunkt bezeichnen, wann Ihre Arbeit zu erscheinen hätte. Wollen Sie demnach, ganz auf das Unbestimmte hin, sich immer schon damit befassen, so gebe ich Ihnen für heute nur noch die

view of the internal evidence which proves on examination
to be insufficiently global. Elsewhere circumstantial evidence
has been accepted which leaves too many other factors out of
account.

The hazards of proceeding without a sound bibliography
can readily be imagined. The traps that await the unwary
include ascribing influence to books which Wagner very
probably had not come across,[66] when the same material was
freely available to him from works he had at home, or suggesting as Wagner's source something he did indeed read, but
not until after he had written the passage in question. With
only Wagner's list and their own hunches to go on, scholars'
work to date has inevitably suffered from distortions and
inaccuracies, so much so that when Wolfgang Golther
achieves such accurate results one is left wondering whether
his success is due to remarkable intuition or sheer good luck,
or whether indeed he had access to more evidence than he
chose to reveal. Equally, the occasional serious defect in Hermann Schneider's paper shows a less fortunate rather than
less conscientious scholar.

Our first task therefore is to establish as accurately as we
can which of the fruits of contemporary Nibelung interest
Wagner knew and, wherever possible, at what date. There is
in fact a quite considerable body of evidence for Wagner's
Ring studies, much of it documentary, some unpublished,
most still unexploited.

There is, first of all, the library Wagner gathered for himself in Dresden. It still exists: after his flight and exile the
library passed into the archives of the Brockhaus family,
Wagner's publishing in-laws, where it remained until Curt
von Westernhagen produced his catalogue of its contents in
1966.[67] The catalogue appeared in time for Deryck Cooke to
forage out Wagner's Edda editions, but it has not yet been

Quellen an, deren Studium mich seiner Zeit für meinen Gegenstand reifte.' (Letter
to Franz Müller, 9 Jan. 1856, *Dokumente*, p. 19.) Müller's book, *Der Ring des
Nibelungen, eine Studie zur Einführung in die gleichnamige Dichtung Richard Wagners*,
eventually appeared in Leipzig in 1862.

[66] It is of course only rarely, if at all, possible to prove conclusively that Wagner
did *not* know a given work provided it was published by the appropriate date.

[67] In *ML*, p. 274, Wagner describes how the loss of his library came about. It is
now in the Richard-Wagner-Museum at Wahnfried.

systematically explored for evidence of Wagner's Nibelung reading.

Secondly, we know from Wagner's autobiography and correspondence that he was a user of the Royal Library at Dresden, the former Königliche Öffentliche Bibliothek, now part of the Sächsische Landesbibliothek.[68] Many of the library's holdings were of course destroyed in the last war, but the catalogues from the period in question survive and, even more valuable, so do the loan journals in which the day-by-day lending activity of the library is recorded. So far the library records have apparently been ignored by Wagner scholars.

Finally, when these two sources have been exhausted there still remains a residue of works which we know or strongly suspect Wagner read. His autobiography and correspondence mention books that he apparently did not find in either his own library or the Royal Library at Dresden. There are further works where the internal and circumstantial evidence point overwhelmingly to their influence on the *Ring* poem, or at least strongly enough to merit consideration. These form a third group to examine.

Our study opens with an exploration of each of these areas in turn to build up so far as possible a trustworthy catalogue of Wagner's *Ring* preparations from among the available Nibelung material of the day.

---

[68] Wagner refers to using books from the Royal Library in his letter to Uhlig of 2 Nov. 1851: *Dokumente*, pp. 57–8; also in his autobiography, e.g. *ML*, p. 390.

PART I

2

Wagner's Personal Library at Dresden

THE library Wagner acquired for himself in the autumn of 1843 in anticipation of a prosperous and settled future was, along with his Breitkopf und Härtel piano and the Cornelius Nibelung print, one of his three most prized possessions in his new Dresden establishment.[1] In his autobiography Wagner relates that while setting up home in the Ostra-Allee he purchased his library 'at once and at one go, proceeding completely systematically following the plan of my intended studies'.[2] Strictly speaking, one might quibble, this cannot be altogether true, since certain of the works contained in his library were not yet published in the autumn of 1843, and there will be further evidence to suggest that some books were later acquisitions.

Nevertheless, Wagner's library scheme was evidently right from the start a characteristically grand-scale undertaking. It eventually numbered almost 200 titles. Of these Curt von Westernhagen found 169 still together when he compiled the library catalogue in the 1960s. The remaining titles are vouched for in a copy of a list left by Minna Wagner of the library contents. Naturally, the accuracy of the bibliographical detail in this list is hard to verify, and Westernhagen has included them cautiously in a kind of appendix to the main catalogue, along with a note on the reliability or otherwise of the information contained.[3]

Several of Wagner's most important Nibelung sources were in his library. We shall start with the works on the list

---

[1] ML, p. 274.
[2] 'sofort, nach dem Plane der mir vorgesetzten Studien durchaus systematisch verfahrend, auf einmal' (ML, p. 274).
[3] DB, pp. 111–13.

he sent to Müller, and then proceed to discover what else his library contained that was of interest to the *Ring*. Westernhagen's catalogue will furnish the basis of this chapter. Titles found in the catalogue proper will be referred to by their catalogue (*DB*) number, while for those in the appendix we shall give the page number.

Wagner's list to Müller began with '*Der Nibelunge Noth und Klage*, edited by Lachmann'. It may surprise us to find the *Nibelungenlied* heading Wagner's list of sources in view of the comment he made in the 'Mitteilung' on the work's dramatic unsuitability (see Chapter 1). A further surprise on perusing the catalogue is that while Wagner possessed four different editions of the *Nibelungenlied*, Lachmann's was not among them.[4] The nearest Wagner owned to a comparable critical edition was A. J. Vollmer's *Der Nibelunge Nôt und diu Klage* (Dichtungen des deutschen Mittelalters, vol. i, Leipzig, 1843; *DB* 99). Like Lachmann, Vollmer based his text on the 'A' manuscript. His introduction is particularly useful, containing among other things a brief resumé of current Nibelung criticism and saga variants.

Lachmann's text, in fact, provided the starting-point for Karl Simrock's perennially popular modern German version, *Das Nibelungenlied*,[5] of which Wagner had the third edition (Stuttgart and Tübingen, 1843; *DB* 101). This formed volume ii of Simrock's six-volume *Heldenbuch*. Wagner nourished quite an admiration for Simrock's modern-language versions of German medieval literature; in *Mein Leben* he speaks of Simrock as 'the saga researcher and reviver I so much admire',[6] and as late as 1878 he was praising the language of Simrock's *Heldenbuch*.[7]

A different manuscript, 'C', supplied the text for *Daz ist der Nibelunge Liet*, of an unnamed editor (Leipzig, 1840; *DB*

---

[4] Westernhagen's comment on Lachmann's edn., 'in DB später fehlend' ('later missing from his Dresden library: *DB*, p. 29) suggests that Wagner may once have possessed this edn. as well; but he gives no evidence in support beyond the appearance of Lachmann's edn. on Wagner's list to Müller.

[5] Or at least for the first 9 edns. In later edns. the text diverged somewhat from Lachmann's.

[6] 'der so sehr von mir geschätzte Sagen-Forscher und Erneuerer' (*ML*, p. 314).

[7] According to Glasenapp: *Dokumente*, p. 19.

98). Neither this nor Gustav Pfizer's modern-German *Der Nibelungen Noth* (Stuttgart and Tübingen, 1843; *DB* 100) pretends to be a critical edition, and probably the text in each case was the least important aspect for Wagner. Their great attraction lay in the woodcuts which illustrated them, particularly those by Julius Schnorr von Carolsfeld and Eugen Neureuther for Pfizer's edition.

In his autobiography Wagner relates that Julius Schnorr arrived in Dresden from Munich to take up the post of director of the gallery. Around the winter of 1845–6 he was to be found among the members of the 'Kränzchen', a circle of Dresden artists who met together on a regular basis for social gatherings. Wagner finds room in his autobiography for some typically deprecating remarks about the personal impression Schnorr created, but at the same time admits that he knew his ancient German sagas. Wagner also writes how struck he had been by some of Schnorr's work: 'I had previously seen some mighty-looking cartoons by him which impressed me very much, both by their size and by the subjects from ancient German history they portrayed, which were of great relevance to me at the time.'[8] Presumably it was the work of Schnorr and Neureuther that prompted Wagner to add the Pfizer edition to his *Nibelungenlied* collection.

Next on Wagner's list is Karl Lachmann's *Zu den Nibelungen und zur Klage* (Berlin, 1836; *DB* 78). The bulk of Lachmann's book, which concerns itself with variant readings in the different manuscripts of the *Nibelungenlied* and like matters, cannot have provided much inspiration for Wagner, but at the end Lachmann includes a reprint of a paper which originally appeared in the *Rheinisches Museum für Philologie*, his 'Kritik der Sage von den Nibelungen'. Lachmann's 'Kritik' provides an imaginative anatomy of the saga as he attempts in three stages to reduce it to its minimum discernible form.

Jacob Grimm's *Deutsche Mythologie* researches with true

---

[8] 'Von diesem [Schnorr] hatte ich zuvor gewaltig sich ausnehmende Kartons gesehen, die mir sowohl durch ihre Dimensionen als durch die damals mir sehr naheliegenden Gegenstände der altdeutschen Geschichte, welche sie darstellten, sehr imponierten.' (*ML*, p. 333).

Teutonic thoroughness the ancient religion and beliefs of the German peoples, with assistance from the better-documented Nordic tradition where native German sources failed. He deals with gods, heroes, and lesser mythological beings, spells and magic, their cosmography and the elements, seasons, flora, and fauna in folk belief. Grimm scrutinizes etymology, place-names, and personal names for his evidence, ritual and superstition, the prose and poetry of the Middle Ages, and the oral traditions still surviving. His finished *Mythologie* is over 1,200 pages long,[9] brimming with detail, and packed with quotations in a dozen or so different languages. Of all the major influences on the *Ring* this was probably the most challenging, and Wagner initially found Grimm's archaeological reconstruction of the nation's lost religion somewhat daunting.[10] He persevered, however, and Grimm's *Mythologie* soon became the 'ever more intimate guide' of his Nibelung studies.[11]

Wagner possessed the two-volume second edition of Grimm's *Mythologie*, published in Göttingen in 1844 (*DB* 44). Evidently, then, this was not among the books Wagner bought 'at once and at one go' on his removal to the Ostra-Allee in the autumn of 1843 (see above), but was one of a number of later additions to his library. That first vivid encounter with the *Mythologie*, which Wagner describes with such warmth in his autobiography,[12] is placed in the summer of 1843, the year before the second edition appeared. Either Wagner's memory was deceiving him as to the exact year of his first reading of Grimm (which would not be completely out of character),[13] or else, as Westernhagen suggests, he read the first edition prior to purchasing the second.[14] In this case we can assume that Wagner either bought the second edition to replace the first edition he already possessed, or that he had borrowed the first edition from some other source.

Wagner's Edda section included both the current translations of the *Snorra Edda*, Christian Friedrich Rühs's *Die Edda*

---

[9] 2nd edn.
[10] *ML*, p. 273.
[11] 'immer vertrauter gewordener Führer' (*ML*, p. 356).
[12] *ML*, p. 273.
[13] See ch. 3 on the *Völsunga saga* (*Vs*).
[14] *DB*, p. 31. The 1st edn. was pub. in Göttingen in 1835.

(Berlin, 1812; *DB* 119) and Friedrich Majer's *Mythologische Dichtungen und Lieder der Skandinavier* (Leipzig, 1818; *DB* 28). This was as well, for Majer intended the 'mythological' in his title quite literally and omitted the 'heroic' sections of *Skáldskaparmál*, including those concerned with the Nibelungs. However, Rühs's edition contained the complete story, and he also translated the verse passages with which Snorri intersperses his account *in situ*. Majer left out Snorri's verse quotations, but included as compensation complete translations of seven items from the *Poetic Edda*, *Vafthrudhnismál*, *Grimnismál*, *Skirnisför*, *Vegtamskvidha*, *Thrymskvidha*, *Hymiskvidha*, and *Völuspá*.

Wagner had a second translation of the *Völuspá*, that by Ludwig Ettmüller, the Germanic philologist he was later to meet in Zurich. Ettmüller's *Vaulu-Spá* (Leipzig, 1830; *DB* 149) contains the text of the poem both in the original and in German translation, an introduction, copious notes, appendices, and a glossary.

Majer and Ettmüller were responsible for the 'Götterlieder' in Wagner's library. The 'Heldenlieder' were apparently supplied by the Grimm brothers. The book had disappeared from the collection in the Brockhaus archives, but the Grimm brothers' *Die Lieder der alten Edda* (Berlin, 1815) is mentioned on Minna's list.[15] Their *Edda* is set out as a parallel-text edition, with Old Icelandic and German translation on facing pages. It contains the 'Heldenlieder' cycle from *Völundarkvidha* through to *Helreidh Brynhildar*; with the exception of the first and second *Gudhrúnarkvidhur* this includes all the poems about the part of the Nibelung saga on which Wagner drew for his *Ring*, and also the Eddic version of the Wieland story, another favourite with Wagner in this period.

Following the poems, the Grimm brothers provide a prose reduction of the events, no doubt to facilitate comprehension of the sometimes obscure and allusive style of the poetry. Their prose reduction sticks to the appointed text with great fidelity; even so, it is remarkable how the slightly greater liberty of the story-teller over the translator, particularly in

---

[15] *DB*, p. 112.

the rendering of names, can add a new interpretative dimension to the Eddic poems.

The Grimm brothers' translation apart, Wagner also had Friedrich Heinrich von der Hagen's complete 'Heldenlieder', not in the translation of 1814 but the original-language edition, *Lieder der älteren oder Sämundischen Edda* (Berlin, 1812; *DB* 27). The obvious question with von der Hagen's edition is how much use Wagner could make of it. In his autobiography Wagner tells that he was prompted by Mone's *Untersuchungen* to read the Edda, 'as far as I was able without a fluent knowledge of Scandinavian languages'.[16] This does not necessarily mean that he made no attempt to read the Norse. Our best evidence for Wagner's practice comes from his copy of the Ettmüller *Vaulu-Spá*, where the German translation follows on from the Old Icelandic text and at the back is a glossary. According to Westernhagen, all three sections are well thumbed, suggesting that Wagner tried to negotiate his way through the original with the assistance available.[17] Given Ettmüller's glossary and the Grimm translation as a crib, Wagner may well have sampled the poems in von der Hagen's Old Norse edition for the flavour of their original language.

Neither the *Völsunga saga* nor its companion in von der Hagen's *Nordische Heldenromane*, the 'Wilkina- und Niflungasaga', or *Thidreks saga*, was in Wagner's personal library. He later wrote that he had been anxious to buy the *Völsunga saga* but could not find it in the Dresden bookshops.[18]

The *Heldenbuch*, a printed collection of German heroic poems, had been appearing ever since the early years of the sixteenth century in various redactions. Nineteenth-century editors approved the concept, but did not necessarily feel bound to confine themselves strictly to the poems of the sixteenth-century *Heldenbuch*, and the various editions of Wagner's day differ substantially in style, language, and content.

[16] 'soweit mir dies ohne fliessende Kenntnis der nordischen Sprachen möglich war' (*ML*, p. 357).
[17] *DB*, p. 36.
[18] Letter to Theodor Uhlig, 12 Nov. 1852, *Dokumente*, pp. 57–8.

Under his *Heldenbuch* entry Wagner had specified: 'old edition; also modernized by Hagen. — Adapted in 6 volumes by Simrock.' When Wagner's entry is unravelled we find three separate editions, and all were in his library. The last-mentioned, 'adapted by Simrock', was Karl Simrock's modern-German *Das Heldenbuch* (6 vols., Stuttgart and Tübingen, 1843–9), which had already been referred to (see Chapter 1). Volume ii was the *Nibelungenlied*, and has also already been mentioned (see above). Volume i, *Gudrun* (*DB* 76), was the least likely to influence Wagner's Nibelung drama. The third volume, *Das kleine Heldenbuch* (*DB* 59), contains a number of relatively short items: *Walter und Hildegunde*, *Alphart*, *Der hörnerne Siegfried*, *Der Rosengarten*, *Das Hildebrandslied*, and *Ortnit*, of which the third, *Das Lied vom Hürnen Seyfrid*, was another major German source of the Nibelung saga.

Those three volumes of Simrock's *Heldenbuch* are entered into the catalogue of Wagner's library under separate headings. The final three volumes comprise Simrock's *Amelungenlied*, and they are catalogued under this title (*DB* 3). The *Amelungenlied* is something of an anomaly among the *Heldenbuch* poems, for here Simrock departed from his usual practice of updating the language of existing poems and instead wrote a new one. His aim was to cover all the remaining areas of the German heroic saga;[19] his justification was that all heroic saga had originally been in poetic form, and that therefore where worthy originals were no longer extant they should be recreated.[20]

Simrock's main source for the *Amelungenlied* was the conglomeration of sagas which form the *Thidreks saga*. Into this he worked existing 'Heldenlieder', fragments of Norse mythology, some folk saga and fairy-tale, and a sprinkling of folk wit. Simrock's determination to include everything leads to a certain amount of awkwardness, and at points the *Amelungenlied* becomes quite tangled, but it is nevertheless a real *tour de force* by the scholar whose works as 'saga reviver' Wagner so esteemed.[21]

---

[19] With the exception of *Wolfdietrich*, which was not included until later edns. of *Das Kleine Heldenbuch*.

[20] Simrock, *Heldenbuch*, vi. 398–9.

[21] *ML*, p. 314; see also above, n. 6.

As the title suggests, the *Amelungenlied* centres loosely around Dietrich von Bern and the exploits of the Amelungs, the heroes at his court. However, the poem starts back in the days of Wieland, and this first section was probably the most interesting from Wagner's point of view. Simrock had in fact earlier published his *Wieland der Schmied* separately (Bonn, 1835), and Golther suggests that it was this Wagner read.[22] However, there is no reason to suppose that Wagner used the 1835 edition either in addition to or instead of the edition he had on his own shelves.

The last volume of the *Amelungenlied* appeared in 1849, the year Wagner left Dresden, though from Westernhagen's catalogue entry it would appear that Wagner was able to acquire it before he fled. All the other volumes of Simrock's *Heldenbuch* had been published by 1846.

Gudrun, the *Nibelungenlied*, *Das kleine Heldenbuch*, *Das Amelungenlied*: all six volumes of Simrock's *Heldenbuch* were present in Wagner's library, though they appear catalogued under their separate volume titles rather than the title of the full series. Since they are entered in the main catalogue they were evidently all still present when Westernhagen produced his catalogue. It comes as a surprise, therefore, to find Simrock's *Heldenbuch* mentioned again in the appendix to the main catalogue, among the books which Minna recorded in the library but which had in the interim disappeared, and even more surprising to read the comment 'only one volume of this was present.'[23] Perhaps the method of entering the individual volumes in the catalogue has led to the presence of the whole series being overlooked; the entry in the appendix remains in any case a mystery.

Simrock's modern–German adaptation is the last of the three *Heldenbuch* editions that Wagner mentions in his letter to Müller. There was also the 'old edition' and von der Hagen's 'modernized' one. 'Old edition' in this context presumably means one in the original language; in this case, of the two remaining *Heldenbuch* editions in his library, Wagner is thinking of *Der Helden Buch in der Ursprache* (Berlin, 1820–5). Edited by F. H. von der Hagen and

---

[22] Golther, *Grundlagen*, p. 12.
[23] 'Hiervon war nur ein Band vorhanden.' (*DB*, p. 112).

A. Primisser, it formed volume ii of the series 'Deutsche Gedichte des Mittelalters' under the general editorship of von der Hagen and J. G. G. Büsching, and Westernhagen catalogues it under this series title, not under the title of *Heldenbuch* (*DB* 21). The volume was published in two parts. The first, appearing in 1820, contains *Gudrun, Biterolf und Dietlieb*, and *Der grosse Rosengarten* and in addition *Otnit* and *Wolfdietrich* from *Kasper von der Rön's Heldenbuch*. The second part contains the rest of *Kasper von der Rön's Heldenbuch* and also *Hörnen Siegfried, Dietrichs Ahnen und Flucht zu den Heunen*, and *Die Ravenna Schlacht*. Part 2 appeared in 1825.

The 'modernized' edition which Wagner also mentions, F. H. von der Hagen's modern-German *Der Helden Buch*, vol. i (Berlin, 1811; *DB* 58), actually appeared before the 'old edition'. It, too, contains the Nibelung source poem *Hörnen Siegfried*, and with it *Das Rosengarten Lied, Etzels Hofhaltung, Alpharts Tod, Ecken Ausfahrt*, and *Riese Siegenot*.

From *Heldenbuch* to heroic saga: Wilhelm Grimm's *Die deutsche Heldensage* (Göttingen, 1829; *DB* 47) is a gathering-together of references to the heroic sagas in literature, and may well have drawn Wagner's attention to sources he was not familiar with. As with Lachmann's *Zu den Nibelungen*, however, the most valuable part of Wilhelm Grimm's *Heldensage* was probably the end, in this case the dissection of the Nibelung saga in the closing essay, 'Ursprung und Fortbildung'.

Franz Joseph Mone's *Untersuchungen zur Geschichte der teutschen Heldensage* (Quedlinburg, 1836; *DB* 93) also shows much evidence of laborious compilation of references. The first three sections, in fact, are given over to the search for saga evidence in the place-names and personal names to be found in historical works and ancient records, together with Mone's interpretation of his evidence. There follows a section of commentary and criticism on the *Nibelungenlied*, the *Poetic Edda* 'Heldenlieder', and *Beowulf*, and the whole is rounded off by studies of individual topics: 'Ueber den Ages', 'Der Hort des Nibelungen', and 'Einfluss des Hortes auf die Bildung'.

Wagner commented on his list to Müller that Mone's *Untersuchungen* had been very important for his Nibelung drama, though no one has been quite sure why. Westernhagen concludes resignedly: 'It may have been a question of an association we can no longer fathom.'[24] We have of course Wagner's evidence that it was Mone's *Untersuchungen*, presumably the central section, which had led him to attempt the Edda (see above). This alone is quite a deserving achievement, and we shall in the course of our study find other things to Mone's credit.

The final work on Wagner's list is Snorri Sturluson's *Heimskringla*, of which he preferred the more readable translation by G. Mohnike to Ferdinand Wachter's tortuously literal one. In his library were both: Mohnike's *Heimskringla*, volume i (Stralsund, 1837; *DB* 133) and Wachter's *Snorri Sturluson's Weltkreis* (2 vols, Leipzig, 1835; *DB* 132). The *Heimskringla* is a strange work to find on a list of *Ring* sources, since apart from the opening chapters of the *Ynglinga saga* it contains nothing on the plot and little on the characters of Wagner's drama. However, its portrayal of Odin-fixated Scandinavian heroic society was useful background, especially for *Die Walküre*, and Wagner found good use for the *Hákonarmál* it contains.

Wagner's personal library at Dresden offered something for eight out of the ten titles on the list to Müller, all apart from the *Völsunga saga* and *Thidreks saga*. There were other works in his library which Wagner did not mention as *Ring* sources but which we shall. They are quite a miscellany. In some cases their claim to inclusion in a list of Nibelung literature may seem rather stretched; we mention Friedrich von Raumer's *Geschichte der Hohenstaufen und ihrer Zeit* (2nd edition, 6 vols., Leipzig, 1840–1), of which Wagner had volumes i, ii, and iv (*DB* 112), only because Wagner was absorbed in the *Hohenstaufen* over much the same period as his Nibelung involvement and, according to *Die Wibelungen* and the 'Mitteilung', saw an almost mystical affinity between Friedrich and Siegfried.

Equally, Ludwig Tieck's 'Gestiefelter Kater', while fairy-tale-based, hardly comes within the circle of Nibelung

[24] 'Es mag an einer für uns nicht mehr greifbaren Assoziation gelegen haben.' (*DB*, p. 35.)

literature, though it is none the less held to be the inspiration behind the scene where Alberich is captured in *Rheingold*.[25] Tieck's 'Gestiefelter Kater' comes in volume v of his twenty-eight-volume *Schriften* (Berlin, 1821–54), of which Wagner had the first twenty volumes (*DB* 143). Wagner's respect for Tieck probably dates back to the influence of his uncle Adolf Wagner, who counted among Tieck's friends;[26] Wagner's own meeting with Tieck in Berlin seems to have been a harmonious occasion.[27]

With the Grimm brothers' *Märchen* we are on firmer Nibelung territory, since they had troubled to establish the connection between their tales and both the Nibelung saga and Germanic mythology (see Chapter 1). The copy of the Grimm brothers' *Kinder- und Haus Märchen* in Wagner's library was apparently the three-volume second edition (Berlin, 1819–22), though as the *Märchen* appear only in the catalogue appendix (p. 112) these details cannot be guaranteed.

Wagner also owned the Grimm brothers' *Deutsche Sagen* in two volumes (Berlin, 1816–18; *DB* 23), and two of Jacob Grimm's technical heavyweights, the *Geschichte der deutschen Sprache* (2 vols., Leipzig, 1848; *DB* 45) and the first part of his *Deutsche Grammatik* (*DB* 43).[28]

How seriously Wagner took his studies of philology and Germanic antiquity can be seen from his holdings of the leading specialist journal, the *Zeitschrift für deutsches Alterthum*. The first six volumes (Leipzig, 1841–8), bound, are mentioned in the library catalogue (*DB* 168). According to Westernhagen, Wagner was a regular subscriber.[29] Assuming he did not cancel his subscription abruptly at the end of 1848, he would still have received before his May departure the first issue of 1849, which opens with Karl Weinhold's article 'Die Sagen von Loki' (pp. 1–94), a comprehensive study of

---

[25] Cooke, *World*, p. 217 n, attributes the Puss-in-Boots influence in *Rheingold* to Perrault's version. While Perrault was the ultimate source of 19th-cent. German retellings, there is no reason to suppose it was Perrault rather than Tieck that Wagner read.
[26] *ML*, p. 30.
[27] *ML*, pp. 359–60.
[28] 3rd edn., pt. 1 (Göttengen, 1840).
[29] *DB*, p. 35.

what became one of the most fascinating characters of the *Ring*.

Everything Wagner needed to know about valkyries was to be found in Ludwig Frauer's monograph *Die Walkyrien der skandinavisch-germanischen Götter- und Heldensage* (Weimar, 1846), listed in the catalogue appendix (p. 112). Based conscientiously on primary source material, Frauer's study is thorough and would have provided Wagner with secondhand access to material he was otherwise barred from, either because it was unobtainable or because of language difficulties. At the end of September 1848, according to Freytag, Wagner expressed himself to be engaged on the subject of valkyries;[30] perhaps he was currently reading Frauer's book.

Despite its title, C. Russwurm's *Nordische Sagen der deutschen Jugend erzählt* (Leipzig, 1842; *DB* 121) is not particularly a book for children. Selected sagas are retold in simplified but non-juvenile form, and the text is followed by a massive critical appendix. Assuming this was one of Wagner's earlier acquisitions, it would have provided a useful introduction to the sagas. Russwurm does not mention Siegfried or the Wälsungs in his selection, but he does start with the Norse gods and includes what may have been Wagner's main text of the Ragnar Lodbrok saga.

Wagner had a copy of Ludwig Uhland's *Gedichte* (Stuttgart and Tübingen, 1842; *DB* 145). He would therefore presumably have been acquainted with the senior poet's short poem 'Siegfrieds Schwert', which appears on page 402.

The *Gedichte* which Anton Alexander Graf von Auersperg published under the pseudonym Anastasius Grün (Leipzig, 1843) were in Wagner's library (*DB* 48). The volume did not, alas, include his *Nibelungen im Frack*, which was published separately the same year. It would be interesting to know whether Wagner ever saw Auersperg's satirical Nibelung poem; it was not the kind of work to be found in the Royal Library.

Hans Sachs's works had enjoyed something of a revival, and Wagner had the first two of the three-volume *Hans Sachs ernstliche Trauerspiele . . . und Possen*, edited by J. G. Büsching

[30] W. Ashton Ellis, *Life of Richard Wagner*, 6 vols. (London, 1900–8), ii. 262, 276. Ellis's *Life* is an expanded trans. of Glasenapp's biography.

(Nuremberg, 1816–24; *DB* 122). Volume ii contained his *Der hörnen Seufried, ein Sohn Königs Siegmund im Niederland*. The play is unlikely to have exercised much independent influence on the *Ring*, but it is satisfying to know that Wagner knew the Siegfried drama of the Mastersinger he revered.

Finally there was one work in Wagner's library which might well have been relevant to his Nibelung drama but which he probably never read. *Saxonis Grammatici Historia Danica*, edited by P. E. Müller (2 vols., Copenhagen, 1839; *DB* 125) sheds some intriguing light on Odin/Wotan and other divinities and has some valuable passages on valkyries. The language, however, was likely to prove an insurmountable obstacle. By his own admission Wagner's Greek had been better than his Latin during his school years,[31] yet when he wished to resume his classical studies in adult life he found he had to resign himself to reading Greek texts in translation, the original language being by then so much beyond his reach.[32] It is unlikely, therefore, that Wagner would have made much headway with the Latin of Saxo, whose reputation for linguistic difficulty had been established centuries previously.[33] Wagner did confess in his autobiography to a certain degree of optimism, or wishful thinking, when purchasing his library, laying in store works in foreign languages he hoped he might some day be able to tackle.[34] Saxo's information on the valkyries would have come to Wagner via Frauer's monograph; the rest Wagner probably had to do without. The *Historia Danica* is a case where the tardiness of Wagner's contemporaries in producing a translation proved a decisive factor in his knowledge of source material.

---

[31] *ML*, pp. 20–2.

[32] *ML*, pp. 221, 274.

[33] In the introd. to Eng. trans. of Saxo Grammaticus, *The First Nine Books of the Danish History* (London, 1894), Oliver Elton quotes the verdict of the author of the epitome from *c.* 1430: 'Since Saxo's work is in many places diffuse, and many things are said more for ornament than for historical truth, and moreover his style is too obscure on account of the number of terms and sundry poems, which are unfamiliar to modern times, this opuscle puts in clear words the more notable of the deeds there related . . .' (p. xvi).

[34] *ML*, p. 274.

3

The Royal Library at Dresden

As soon as he had settled permanently in Dresden and while he was establishing his own private library, Wagner started to frequent Dresden's leading public lending library, the Königliche Öffentliche Bibliothek, or Royal Library. The books he sought were on similar subjects to those he was buying for his own library. Wagner was acquainted with a librarian there who included philology among his specialisms, Johann Georg Theodor Grässe, and it had been suggested that Grässe was able to direct Wagner's attention to works which would interest him.[1] Wagner's autobiography ignores Grässe completely. Nevertheless, Grässe's *Gesta Romanorum* translation was in Wagner's library, (*DB* 35),[2] and when the librarian published his *Sage vom Ritter Tanhäuser* in 1846 it was dedicated to 'his dear friend, the Royal Saxonian court conductor Richard Wagner'.[3] Evidently there was an acquaintanceship of some cordiality between the two.

The loan journals of the Royal Library for this period were kept on a yearly basis. Each annual volume is divided into sections alphabetically, and works were entered into the appropriate section as they were borrowed according to the name of the author or other heading. The pages each consist of several columns, providing for author and short title, name of borrower, date of loan, and date of return.

For establishing a bibliography of Wagner's *Ring* reading, the loan journals are obviously of immense importance, since they contain not only a record of all the books Wagner bor-

[1] *DB*, p. 17; Ellis, *Life*, ii. 99. Apart from his philological interests Grässe published a highly successful series of guides to the royal porcelain and other collections at Desden.

[2] 2 vols. (Dresden and Leipzig, 1842).

[3] 'seinem theuern Freunde, dem königl. Sächs. Hofkapellmeister Richard Wagner' (*Die Sage vom Ritter Tanhäuser, aus dem Munde des Volks erzählt* (Dresden and Leipzig, 1846)). In the pref. Grässe refers to Wagner's opera.

rowed but also the sometimes crucial date of his reading. Nevertheless, invaluable as they are, in two areas the loan journals present us with tantalizing incertitude. The practice of entering short titles only, and no date of publication, makes it difficult, where more than one edition of a book had been published, to know which one the entry refers to. The holdings of the Royal Library were admirably comprehensive and offered, for instance, several of von der Hagen's *Nibelungenlied* editions. Worse still, the short titles may be ambiguous and leave much doubt as to whether, in this particular instance, the *Nibelungenlied* is meant at all. Fortunately, ambiguity on such a scale occurs only once.

A more serious problem is presented by the system of entering borrowers' surnames only. Wagner is not an uncommon name, and there is no reason to assume that every time 'Wagner' appears in the borrower column of the loan journals it refers to the Kapellmeister. It would, for instance, be difficult to imagine what 'our' Wagner was doing borrowing Rienzi's *L'Océanie*, and it is hard to believe, in view of what we have already established about Wagner's command of Latin, that it was he who had out on loan an almost continuous flow of Latin classics in the original language. In any case, the Roman authors continued to be issued to 'Wagner' long after Richard had fled Dresden.

Obviously there was more than one Wagner borrowing from the Royal Library in that period, and the loan journals are no help to establish which. Nevertheless, we can take heart from the fact that loans to 'Wagner' drop off remarkably after early May 1849. It is also encouraging to note that large quantities of Nibelung literature were being borrowed from June 1848 onwards, when Wagner's involvement with the *Ring* material was reaching its first peak. We have at least one fixed point on which to orientate ourselves, since Wagner specifically mentions borrowing the *Völsunga saga* from the Royal Library.[4] If we proceed with reasonable caution, correlating dates and taking due account of Wagner's periods of absence from Dresden, we shall probably not go too far astray if we assume that works likely to be of special

---

[4] Letter to Theodor Uhlig, 12 Nov. 1851, *Dokumente*, pp. 57–8.

interest to the author of the *Ring* were in fact borrowed by him and not by some unknown namesake.

The first clutch of Nibelung literature appears in the loan journals around 1844–5. The mystery of Lachmann's *Nibelungenlied* edition, which Wagner cites in his list to Müller but which was not among those in his own library, is here resolved, for he borrowed it from the Royal Library on 1 February 1845, returning it a month later.[5]

On the same day that the Lachmann *Nibelungenlied* was issued to Wagner, Joseph von Hinsberg's modern-German version, *Das Lied der Nibelungen* (4th edition, Munich, 1838), was returned to the library by a Herr Wegner, who had had it out on loan since 11 January the same year. This seems to have been Herr Wegner's only visit to the Royal Library, for there is no further mention of his name in the loan journals of the period. There was indeed a Herr Wegener who was also borrowing books around this time; but the return of the Hinsberg *Nibelungenlied* on the same day that Wagner visited the library to borrow Lachmann's edition does foster the suspicion that the solitary 'Wegner' is in fact our Wagner, all the more so since we shall soon find that Wagner was quite in the habit of borrowing different editions of the same work simultaneously or in quick succession. There is nothing so distinctive about the Hinsberg *Nibelungenlied* that Wagner's drama stood to gain or lose by his knowing it, but it does include six rather pleasing copperplates.

About a year earlier, on 17 January 1844, Wagner had borrowed the 'Nibelungen von Hagen', which he returned on 18 February. Of all the entries in the loan journals this is the one where the absence of full bibliographical details, or even a full title, is most teasing. Was the book in question that arch-Romantic among all Romantic Nibelung studies, Friedrich Heinrich von der Hagen's *Die Nibelungen, ihre Bedeutung für die Gegenwart und für immer* (Breslau, 1819)?

<hr>

[5] This still does not fully explain why Wagner cited Lachmann's edn. specifically to Müller, for it is unlikely to have made any material difference to the *Ring* which version of the *NL* he used. We can only guess that Lachmann was named either because he was academically the most prestigious editor or because Wagner had been rereading his version recently in Zurich.

This is the work where von der Hagen, abandoning the relative restraint of the editor and translator, strives to show the *Nibelungenlied* as an essentially Christian manifestation of universal myth. Nowhere does von der Hagen give freer rein to the 'muddled erudition' with which Hermann Schneider credits him;[6] possibly the soundest statement in the whole book is the author's admission towards the close: 'I may have made the odd mistake here too, particularly in the mythological interpretation of the Nibelungs attempted above . . .'[7] Nevertheless, one would like to think that Wagner knew of von der Hagen's interpretation of the Nibelung saga as 'the original ancestral saga of the human race itself, of Paradise and the Fall of Man, how through the serpent (with the human head), through woman and through gold sin and death entered into the world. . . .'.[8]

If von der Hagen's study were being entered in the loan journal under a short title, then *Nibelungen* is undoubtedly what that title would be. However, as we earlier remarked, the term *Nibelungen* was the popular form for referring to the *Nibelungenlied*, and the well-stocked Royal Library held a full four of von der Hagen's *Nibelungenlied* editions.[9] From the evidence of the loan journals we cannot tell whether the *Nibelungen* Wagner borrowed was von der Hagen's study or merely another edition of the *Nibelungenlied*.

Henri Lichtenberger and Hermann Schneider have both assumed on internal evidence that Wagner knew the study.[10] The internal evidence, it must be admitted, is not totally compelling: for every instance that these two scholars cite plausible alternative sources can be found elsewhere in Wagner's Nibelung studies. Nevertheless, the overall

[6] 'wirre Belesenheit' ('Altertum', p. 109).

[7] 'Ich mag auch hier im Einzeln wieder irren, zumal in der vorn versuchten mythologischen Erklärung der Nibelungen . . .' (pp. 196–7).

[8] 'die Ur- und Stamm-Sage des Menschengeschlechtes selber, von dem Paradiese und Sündenfalle, wir durch die Schlange (mit dem Menschenkopfe), durch das Weib und das Gold, die Sünde und der Tod in die Welt gekommen . . .' (von der Hagen, *Die Nibelungen*, pp. 66–7).

[9] *Der Nibelungen Lied nebst Glossar* (Berlin, 1807); *Das Nibelungenlied in der Ursprache* (Berlin, 1810); 3rd edn. (Breslau, 1820); *Der Nibelungen Noth*, 3rd edn. (Breslau, 1820).

[10] H. Lichtenberger, *Richard Wagner, poète et penseur*, 3rd edn. (Paris, 1902), p. 267 n; Schneider, 'Altertum', pp. 109 ff.

evidence is sufficiently strong to ensure that von der Hagen's study cannot be ignored in a list of influences on Wagner's Nibelung drama. We shall therefore admit it, with reservations, into our bibliography of Wagner's *Ring* reading, and in the present study at all events reference to von der Hagen's *Die Nibelungen* will indicate his *Die Nibelungen, ihre Bedeutung für die Gegenwart und für immer*.

There followed something of a lull in Wagner's Nibelung borrowings. In the spring of 1847, on 19 April, he borrowed Jacob Grimm's *Deutsche Grammatik*, third edition, part 1 (Göttingen, 1840), which he returned on 7 August. Then on 10 June 1848 Wagner initiated his most intensive phase of *Ring* reading by borrowing the Grimm brothers' *Lieder der alten Edda* (Berlin, 1815) and Ferdinand Wachter's *Snorri Sturluson's Weltkreis* (Leipzig, 1835), which were out on loan to him until 13 September.

The remarkable feature about these three works is that they were all to be found in Wagner's own personal library at Dresden. This, of course, raises the pressing question of why Wagner should borrow books from a public library if they were already on his shelves.

The first, and potentially most alarming, possible answer is that the three works were borrowed not by Richard but by another of the Royal Library's Wagners. Such a solution would not affect the certainty with which the three works in question appear on Wagner's *Ring* list, since that was established in Chapter 2. To discover that another Wagner in Dresden was actively borrowing Nibelung literature from the Royal Library at this period would, however, force us to query other loans of potential *Ring* sources and would certainly upset the confidence with which we can assume that they were made to the court Kapellmeister. At this point it must be stated that some Wagner was borrowing a work known to have interested Richard, Raumer's *Geschichte der Hohenstaufen*, on 11 October 1847, when the *Ring* author was in Berlin. Raumer's work on the Hohenstaufen, as we have already observed (see Chapter 2), is not Nibelung literature except in the eyes of the author of *Die Wibelungen*; nevertheless, it is treading close.

There are, happily, other possible answers. Wagner may have borrowed the books from the Royal Library because he had temporarily mislaid his own copies, or was unable to lay hands of them: the loan of Grimm's *Deutsche Grammatik* coincides with Wagner's removal to the Marcolinisches Palais.[11] We may even conclude, rather cynically, that Wagner preferred to take volumes from the Royal Library away on his travels rather than expose his own books to the attendant risks; the Grimm *Edda* and Wachter's *Heimskringla* may have been borrowed in anticipation of his Vienna trip in July.[12] But perhaps the most satisfactory solution is that when Wagner borrowed the works in question they were not yet in his own library, and that he purchased them only after, and probably as a result of, reading them from the Royal Library.

The flow of Nibelung literature continued throughout the rest of the year. On 21 August 1848 Wagner borrowed for the first time Jacob Grimm's *Deutsche Rechtsalterthümer* and two more Edda translations. After the Grimm 'Heldenlieder' he began to discover more of the 'Götterlieder'. J. L. Studach's *Sämund's Edda des Weisen* (Nuremberg, 1829) contains the *Völuspá* and *Hávamál*, *Vafthrudhnismál*, *Grimnismál*, *Alvíssmál*, *Hymiskvidha*, *Thrymskvidha*, and *Harbardhsliodh*. It was intended as part of a series of Edda translations, and was issued as 'Abtheilung [part] 1', but no more volumes were ever published. At the same time Wagner borrowed Gustav Thormod Legis's 'Fundgruben des alten Nordens', another mysteriously truncated series of which it appears only part 1 of volume ii was ever in circulation. Volume ii, *Edda, die Stammutter der Poesie und der Weisheit des Nordens* (Leipzig, 1829), was planned to cover the 'Götterlieder' in parts 1 and 2 and the 'Heldenlieder' in part 3. Part 1 contains *Völuspá* and *Vafthrudhismál*, *Grimnismál*, *Hymiskvidha*, *Skirnisför* and *Harbardhsliodh*; parts 2 and 3 were apparently never published. Only volume ii, the Edda, is listed as present in the Royal Library holdings according to the catalogues, and the loan journals say that it was part 1 that Wagner borrowed. It was

[11] *ML*, p. 354.
[12] *ML*, p. 380.

evidently therefore the six 'Götterlieder' listed above that Wagner had out on loan.

Wagner was away from Dresden for a few days at the end of August, visiting Liszt in Weimar,[13] but biographical sources are not specific as to which dates were involved. We cannot therefore prove that Wagner was around in Dresden when the *Rechtsalterthümer* and the two Eddas were lent out; equally, there is nothing to indicate that he was away on that particular date, and in the absence of evidence to the contrary we shall assume that it was indeed he who borrowed the three works in question.

Wagner returned all three on 2 October. On 21 October he took the *Rechtsalterthümer* out again, and three other works. One was yet another Edda translation, *Die Lieder der Edda von den Nibelungen* (Zurich, 1837), by Ludwig Ettmüller, or 'Eddamüller' as Wagner is reputed to have called him.[14] Leaving out the Wieland and Helgi poems, Ettmüller's translation starts at *Sigurdharkvidha Fafnisbana fyrsta* and carries on beyond where the Grimm brothers leave off to complete the 'Heldenlieder' cycle, including for good measure a poem entitled *Gunnars Harfenschlag*.

The second item Wagner took out was the Grimm brothers' three-volume *Altdeutsche Wälder* (Kassel and Frankfurt-on-Main, 1813–16), containing scores of articles on medieval language and literature. Probably the *Altdeutsche Wälder* reached Wagner too late to influence his incipient Nibelung drama, but the work must have afforded some retrospective interest.

Finally, Wagner borrowed F. H. von der Hagen's *Nordische Heldenromane*, volume iv (Breslau, 1815), the *Völsunga saga* he had been so anxious to locate. This of all the loans from the Royal Library is the one we can be most sure of, since Wagner wrote to Uhlig that it was here he had eventually located it after much searching:

Already while at Dresden I made every conceivable effort to buy a book which, however, no longer existed anywhere in the bookshops. I found it finally in the Royal Library. It's a slim, small

---

[13] *ML*, p. 385.
[14] Schneider, 'Altertum', p. 115.

octavo or even duodecimo volume and it's called *The Völsunga saga*—translated from the Old Norse by H. von der Hagen.[15] There is no prior reference in the loan journals to the *Völsunga saga* being lent out to Wagner, so we can therefore be sure of the date, too, when he first found the saga. Were we not so certain, we might be inclined to think Wagner had come upon the *Völsunga saga* a good year earlier.

His autobiography tells of pleasant afternoons in the summer of 1847 spent in the grounds of the Marcolinisches Palais, to which he had removed that spring, absorbed in his studies. After the Greek dramatists it was early Germanic literature that held his attention, and apart from the Eddas and Mone's familiar *Untersuchungen* Wagner specifically mentions the *Völsunga saga*: 'Of decisive influence on the treatment of this material which was evolving in me was my reading of the *Völsunga saga*, aided by Mone's *Untersuchungen*.'[16]

Against this summer of 1847 that Wagner so fondly remembers, the loan journals unrelentingly insist on a date in late October 1848 for the first introduction to the *Völsunga saga*. Fascinatingly enough, the evidence of the loan journals is supported on quite other grounds by William Ashton Ellis, who translated and revised Glasenapp's biography of Wagner. 'It is manifest', he writes, 'that in this autumn [1848] fell his reading of the "Völsunga-Saga".' To this he adds a footnote:

Perhaps we can even *date* this discovery: on Nov. 3, 1848, just nine days before commencing to engross the poem of *Siegfried's Tod* in its final Dresden form, Wagner writes to old Fischer that 'owing to a windfall' (durch einen eingetretenen Fall) he is most anxious to have his mornings free. As the morning hours were those he devoted to creative or literary work, whenever possible, it looks very much as if he had just 'happened' on the Völsunga-Saga. Whether we assign his first acquaintance with it to Nov. 3 1848, or

---

[15] 'Schon in Dresden gab ich mir alle erdenkliche Mühe, ein Buch zu kaufen, das aber nirgends im Buchhandel mehr existierte. Ich fand es endlich auf der königlichen Bibliothek. Es ist ein dünnes Kleinoktav oder gar Duodez-Bändchen und heisst: "*Die Wölsungasaga*"—aus dem Altnordischen übersetzt von H. van [i.e., von] der Hagen.' (Letter to Theodor Uhlig, 12 Nov. 1851, *Dokumente*, pp. 57–8.)

[16] 'Von entscheidendem Einfluss auf die bald in mir sich gestaltende Behandlung dieses Stoffes war an der Hand der Moneschen "Untersuchungen" die Lektüre der Wälsungasaga.' (*ML*, p. 357.)

a few weeks earlier, it immediately inspired him with the conception of a great part of *Die Walküre* . . .[17]

Evidently Wagner's memory in *Mein Leben* was racing
ahead of events somewhat. He mentions the Eddas alongside
the *Völsunga saga* among his summer-time reading, and we
have noted that these too belong chiefly to the second half of
1848 rather than 1847. Quite possibly his recollection of the
summer of 1847 was merging with that of the following
year. Outwardly they were very similar; in neither year did
Wagner move out of Dresden for the summer as had been his
habit in the past. One dreamy summer afternoon spent in the
shade of the shrubbery or perched on the statue group of the
dried-up fountain in the grounds of the Marcolinisches Palais
must have seemed very like another in retrospect when
Wagner came to write *Mein Leben*.[18]

Wagner returned the *Völsunga saga* and Ettmüller's *Edda* on
29 January 1849. The *Rechtsalterthümer* and *Altdeutsche Wälder*
did not go back to the Royal Library until 20 June 1849,
doubtless as part of Minna's clearing-up operation. Wagner
maintained his borrowing momentum to the end, gravitating
to the twilight zone between saga and history of his
Wibelungen essay with works such as Göttling's *Über das Geschichtliche im Nibelungenliede* (Rudolstadt, 1814) and Emil
Rückert's *Oberon von Mons und die Pipine von Nivella* (Leipzig,
1836). These he borrowed on 10 February 1849; they found
their way back on 19 June. Neither these nor other of his later
borrowings, however, were to exercise any further influence
on his Nibelung drama.

---

[17] *Life*, ii. 275 and n.
[18] *ML*, pp. 356–7.

4
Wagner's Other *Ring* Reading

ONE thing is certain: Wagner obtained material for his *Ring* from sources other than the Royal Library and his own collection. We can be sure of this if only because Wagner himself mentions two books in his correspondence and autobiography which we have not so far accounted for. One is that earlier (first) edition of Jacob Grimm's *Deutsche Mythologie* (Göttingen, 1835) which he was apparently reading on his summer trip to Teplitz in 1843 (see Chapter 2). The other is the *Thidreks saga*, translated by F. H. von der Hagen and published as volumes i–iii of his *Nordische Heldenromane* (Breslau, 1814), which Wagner mentions in his list to Müller.

Both books were in fact in the Royal Library, though it appears that Wagner did not borrow them from there. Over the *Thidreks saga* in the *Nordische Heldenromane* there may be room for some doubt on this point. The loan journal entry made when Wagner borrowed the *Völsunga saga* reads: 'Hagen, Heldenromane. 4.' It could be argued that the '4.' means 'four volumes', which would include the *Thidreks saga* as well as the *Völsunga saga*. However, this would go against standard German practice, where '4.' could mean only 'fourth', not 'four'. It would also go against the practice of the loan journals which, where more than one volume of a work was being borrowed, list each volume number individually. If Wagner had been borrowing the *Thidreks saga* as well as the *Völsunga saga* on 21 October 1848 we would therefore expect to read: 'Hagen, Heldenromane. 1, 2, 3, 4'. There is in fact a certain amount of internal evidence to suggest that Wagner had read the *Thidreks saga* before this date.

As to what Wagner's alternative sources of supply might have been, there is no very clear answer. There were of course other public libraries in Dresden at that time, foremost the Secundogenitur, but unfortunately the surviving records are not sufficiently complete to afford any real assistance.[1]

Then there were friends, such as the members of Hiller's 'Kränzchen', where Wagner had met Schnorr von Carolsfeld.[2] Admittedly, from Wagner's account they do not sound particularly promising, with the possible exception of the architect Gottfried Semper. Wagner and Semper had been at aesthetic loggerheads until Wagner managed to convince the architect that his real interest was in 'German antiquity and discovering the world of ideas of ancient Germanic myth'.[3] At this Semper softened: 'Once we got on to pre-Christian religion and I informed him of my enthusiasm for heroic saga proper he became quite a different person, and a manifestly deep and serious interest now began . . . to unite us.'[4] Even Wagner's neglected librarian friend Dr Grässe may have helped him to further books.

Back in his Paris days Wagner had of course associated with men from the bibliographical worlds—Lehrs, Anders, Avenarius; it was then that he read Raumer's *Geschichte der Hohenstaufen* for the first time.[5] The major intellectual influence of his boyhood years, regressing still further into the past, was his uncle, the scholar and translator Adolf Wagner, who numbered among his literary friends not only Tieck but also Fouqué, author of the Nibelung trilogy *Der Held des Nordens*.[6]

Wherever and whenever the encounter with Fouqué's Nibelung drama occurred, the internal evidence speaks com-

[1] The library itself is no longer intact and the surviving records do not give a reliable account of the holdings in the 1840s. I am indebted to Dipl. Phil. W. Stein of the Sächsische Landesbibliothek for this information.

[2] *ML*, pp. 332 ff. See ch. 2.

[3] 'das deutsche Altertum und die Auffindung des Ideales des urgermanischen Mythus' (*ML*, p. 334).

[4] 'So wie wir nun in das Heidentum gerieten und ich ihm meinen Enthusiasmus für die eigentliche Heldensage kundgab, ward er ein ganz anderer Mensch, und ein offenbares grosses und ernstes Interesse begann uns jetzt . . . zu vereinigen.' (*ML*, p. 334.)

[5] *ML*, pp. 167, 180–2, 221.

[6] For the importance of Adolf Wagner in Richard's intellectual development see *ML*, pp. 15–16, 29–30. Adolf Wagner's friendship with Fouqué is mentioned by Arthur Drews, *Die Ideengehalt von Richard Wagners dramatischen Dichtungen in Zusammenhang mit seinem Leben und seiner Weltanschauung* (Leipzig, 1931), 101–3.

pellingly for Wagner having known it. The parallels between
the first part of Fouqué's trilogy, *Sigurd der Schlangentödter*,
and Wagner's two Siegfried dramas are numerous, and in
many cases the influence of the elder poet's work is unambiguous. Such instances range from the act structure of *Siegfried*, which closely mirrors the first three sections of
Fouqué's *Sigurd*, through to minor episodes in *Götterdämmerung*, and affect mood, ethos, and the style and tone of the
language. Arthur Drews had pointed to Fouqué's early use of
alliteration in the 'song' sections of *Sigurd*,[7] and Golther has
drawn attention to the similarities of language in the forge
scenes.[8] We shall find further that until the end of October
1848 Wagner relied chiefly on Fouqué for his knowledge of
the *Völsunga saga*. 'Wagner had Fouqué's trilogy right in
front of his eyes when he composed his poem' is Arthur
Drews's verdict,[9] and despite the lack of confirmatory
external evidence Fouqué's *Sigurd* is included unhesitatingly
in our *Ring* bibliography.

Ernst Raupach's *Der Nibelungen-Hort* is the one dramatic
work Wagner admitted knowledge of: 'Before I had my
Nibelung poem printed and distributed at the beginning of
1853 the subject matter of the medieval *Nibelungenlied* had to
my knowledge only once been adapted as a play, by Raupach
in his prosaic manner, and as such had been performed,
without success, in Berlin.'[10] Wagner does not of course say
that he had read the work or seen it performed, but the fact
that he ventures judgement on the piece does suggest first-
hand acquaintance. Possibly he had seen the play during one
of his Berlin trips; one thinks particularly of his longer stays
in 1836 and 1847. Raupach's *Nibelungen-Hort* is, as Wagner
indicates, based too closely on the *Nibelungenlied* to have
influenced the *Ring* composer, except for the prologue,

---

[7] *Ideengehalt*, pp. 101–3.
[8] *Grundlagen*, pp. 64–5.
[9] 'Wagner hat Fouqué's Trilogie bei der Ausarbeitung seiner Dichtung unmit-
telbar vor Augen gehabt.' (*Ideengehalt*, p. 101.)
[10] 'Bevor ich, im Beginne des Jahres 1853, mein Nibelungen-Gedicht drucken
und verteilen hatte lassen, war der Stoff des mittelalterlichen Nibelungenliedes
meines Wissens nur einmal, und zwar bereits vor längerer Zeit, von Raupach in
seiner nüchternen Weise zu einem Theaterstück bearbeitet, und als solches, ohne
Erfolg, in Berlin aufgeführt worden.' (*Schr.*, vi. 261.)

where Raupach's combination of *Das Lied vom Hürnen Seyfrid* and Scandinavian material produces results intriguingly akin to Wagner's own.

At Zurich Wagner had the chance to meet and become acquainted with the Germanist Ludwig Ettmüller, whose translations of *Völuspá* and the Eddic Nibelung poems Wagner already knew. Ettmüller had done a great deal in the field of heroic literature, including translating *Beowulf*. A mutual acquaintance, Frau Wille, reported in the spring of 1852 that Ettmüller had told her Wagner 'was studying the Norse heroic sagas and the Edda, and seeking advice and explanations, for which reason Ettmüller saw him often.'[11]

No doubt Wagner stood to learn a lot from Ettmüller, and various scholars have assumed he did. Deryck Cooke suggests Ettmüller was instrumental in supplying Wagner with texts,[12] while Schneider is almost prepared to credit Ettmüller with having acquainted Wagner with the Eddas singlehanded.[13] Even Golther and Ashton Ellis grudgingly allow some influence from the Zurich philologist.[14] Wagner himself remains characteristically silent on the subject.

It was while Wagner was at Zurich that Simrock's new, complete Edda translation appeared. Wagner writes of Simrock's *Edda* in the 'Epilogischer Bericht', where he mentions 'the poems of the Edda, which Simrock subsequently made very readily accessible . . .'[15] The impression Wagner may be giving here is that Simrock's *Edda* appeared too late to help him; certainly, we have seen and admired Wagner's efforts to familiarize himself with the Eddic poems before Simrock's full translation gave such easy access. All the same, Simrock's *Edda* appeared well before Wagner completed his *Ring* poem and there is no reason to believe that once it was available Wagner abstained from reading it. Indeed, the 'Epilogischer Bericht' would indicate that he *did* know it. The question is, when?

The *Ring* writing of the Zurich period is characterized by a

[11] Ellis, *Life*, iii. 323.
[12] *World*, p. 111 n.
[13] 'Altertum', p. 115.
[14] Golther, *Grundlagen*, pp. 13–14; Ellis, *Life*, iii. 323, 419.
[15] 'die Lieder der *Edda*, welche seitdem durch Simrock sehr leicht zugänglich gemacht worden waren . . .' (*Schr.*, vi. 262).

marked upsurge of direct influence from the *Poetic Edda*, generating fresh scenes, shaping the *dramatis personae*, formulating the dialogue patterns, and creating new moods. It has been reasonably argued that Wagner's *Siegfried* is directly indebted to *Vafthrudhnismál*, *Alvíssmál*, *Fafnismál*, *Vegtamskvidha (Baldrs Draumar)*, *Fiölsvinnsmál (Svipdagsmál)*, and *Sigrdrífumál*;[16] that in *Das Rheingold* the influence of the *Snorra Edda* and the poems *Oegisdrecka (Locasenna)* and *Reginsmál* is particularly strong,[17] while *Die Walküre* shows traces of *Grimnismál* and *Helgakvidha Hjörvardhssonar*;[18] that features of the poem *Hrafnagaldr Odhins*, since rejected as spurious but at the time included in Edda editions, are apparent in the conversion of *Siegfrieds Tod* to *Götterdämmerung*;[19] and that the poem *Völuspá* is in evidence throughout.[20] The linguistic parallels are so close that Wagner must obviously have had very recent access to the texts. Three of the poems — *Fiölsvinnsmál*, *Oegisdrecka*, and *Hrafnagaldr Odhins* — were not in any of the translations which we have examined so far.

At this point the Wagner scholars mentioned above tend to bring in Ettmüller's name to explain the new Eddis spirit of Wagner's Zurich *Ring* writing. Yet even though translations of all the poems listed above had appeared in print, including the three not in Wagner's other editions,[21] it would have severely tried Ettmüller's ingenuity to assemble them all in the spring of 1851 without the help of Simrock's *Edda*. In any case, the little information we have on Wagner sounding out the philologist on his specialism belongs to the spring of 1852, a full year after the *Poetic Edda* had made its impact on Wagner's Young Siegfried drama in May 1851.

The obvious answer is that, whatever Ettmüller's role,

---

[16] Golther, *Grundlagen*, pp. 67, 71, 77, 79, 82; Schneider, 'Altertum', p. 118; Cooke, *World*, pp. 74, 111.

[17] Golther, *Grundlagen*, pp. 30–3; Schneider, 'Altertum', pp. 117, 118; Cooke, *World*, pp. 109, 110, 117.

[18] Golther, *Grundlagen*, pp. 58, 59; Cooke, *World*, p. 110.

[19] Golther, *Grundlagen*, p. 97.

[20] Golther, *Grundlagen*, p. 83; Schneider, 'Altertum', pp. 116, 117; Cooke, *World*, pp. 111, 144–5, 227, 240, 317–18.

[21] *Fiölsvinnsmál* and *Oegisdrecka* trans. F. D. Gräter in *Nordische Blumen*, pp. 172–90, 209–33; *Hrafnagaldr Odhins* trans. Gräter in *Idunna und Hermode*, 4 (1816), Nos. 34, 35, 36, 39.

Wagner was reading Simrock's new Edda translation during the lead-up to his resumption of the *Ring*. It is indeed so obvious that one questions why even those scholars who include Simrock's *Edda* among Wagner's sources hesitate to link it too openly with Wagner's major 1851 creation, the Edda-filled scenes of *Der junge Siegfried*.[22] Perhaps the reluctance stems from the fact that Simrock's *Edda* was published only that same year, and there may be doubts as to whether it was already available when Wagner commenced his Young Siegfried drama. It would of course be imprudent to recommend as a source a work which had not then come out.

However, according to a letter which Simrock's publisher wrote to him on 14 March 1851 his *Edda* had been published and distributed by the end of February,[23] and so Wagner would have had a good two months to read and digest the new translation before resuming work on the *Ring* in May. We therefore conclude that Wagner did know Simrock's *Edda*, and that he first read it between March and May 1851. If we further reflect that Wagner's Nibelung drama had been lying dormant for over two years prior to the appearance of Simrock's *Edda*, and that Eddic influence is so prominent in the work that followed, we may hazard that reading Simrock in the spring of 1851 was instrumental in reawakening Wagner's *Ring* inspiration.

Simrock's *Edda* is the last work where the evidence is so overwhelmingly in favour that it merits inclusion on our list with no reservations. We now turn to the remaining handful, where either the prospective influence on Wagner's Nibelung drama or the likelihood of his knowing it, or both, ranges from the marginally uncertain to the highly doubtful.

The first such work is Friedrich Theodor Vischer's

---

[22] See e.g. Golther, *Grundlagen*, p. 12: 'Simrocks Edda bot ihm aufs bequemste die Göttersagen, die im Rheingold und in den Wotanszenen vorkommen.' ('Simrock's *Edda* presented him with the myths which appear in *Das Rheingold* and in the Wotan scenes in the most convenient form.') Cooke limits himself to saying that Wagner 'could have had a copy sent to him in Zürich.' (*World*, p. 111.)

[23] Letter in the archives of the former Cotta-Verlag, now in the Deutsches Literaturarchiv, Marbach. I am indebted to Dr Jochen Meyer, Director of the Cotta-Archiv, for the information on the publication and distribution of Simrock's *Edda*. The letter to Simrock actually states: 'Die Edda wurde . . . nach allen Theilen Deutschlands versandt . . .' ('The *Edda* was . . . dispatched to all areas of Germany . . .'), but I am assured that 'Deutschland' in this context would include Zurich.

'Vorschlag zu einer Oper'. We have already mentioned this
appeal for a German national opera (see Chapter 1), and the
general consensus is that Wagner knew it.[24] Whether he did
or not probably made little difference. Given the climate of
the age and his own field of interest Wagner would quite
possibly in due course have turned to the Nibelungs anyway
as a suitable opera subject. Vischer's own specimen scenario,
firmly grounded in the *Nibelungenlied*, offers a kind of converse to Wagner's own.

Vischer had more success with the authoress Luise Otto,
who wrote three articles on the subject and published three
scenes of a libretto, all in the *Neue Zeitschrift für Musik*.[25]
Wagner read and occasionally contributed to the journal, and
so may well have seen the three scenes. It was probably Luise
Otto's libretto which was the subject of an offer of collabora-
tion broached to Wagner by Gustav Klemm. Wagner politely
declined the offer ' as a matter of course',[26] and so presumably never became better acquainted with the libretto.

We cannot be certain that Wagner knew Görres's *Der
hürnen Siegfried*, though there are passages in Görres's epi-
logue on the eastern origin of the Germanic peoples and the
Emperor Friedrich as the new Siegfried that put one very
much in mind of Wagner's *Wibelungen* essay. Equally, the
evidence that Wagner had read Göttling's *Nibelungen und
Gibelinen* concerns *Die Wibelungen* rather than the Nibelung
drama. It is, however, unusually strong evidence, and
Schneider, Hans Lebede, and Henri Lichtenberger have all
concluded that Wagner knew the work.[27] Since there is one
passage in *Götterdämmerung* which appears to show Göttling's
distinctive influence we shall include his *Nibelungen und
Gibelinen* on our reserve list.

Wagner rated F. J. Mone's *Untersuchungen* highly (see
Chapter 2), and it is always possible that he took the trouble
to read the same author's earlier work, the *Einleitung in*

[24] Drews, *Ideengehalt*, pp. 101–3; E. Newman, *The Life of Richard Wagner*, 4 vols.
(paperback edn., Cambridge, 1976), ii. 25–7; Cooke, *World*, p. 131.

[25] Vol. 23, Nos. 8, 13, 33, 43, 44, 46 (12 Aug. – 5 Dec 1845).

[26] 'von vornherein' (*Dokumente*, p. 24).

[27] Schneider, 'Altertum', p. 110; Lebede, *Musikdramen*, ii. 9; Lichtenberger,
Richard Wagner, pp. 250 n, 256 n.

das Nibelungen-Lied zum Schul- und Selbstgebrauch bearbeitet
(Heidelberg, 1818). Mone's mythological interpretation of
the saga in the second half of the *Einleitung* would have been
most interesting to Wagner and had a lot to offer the final act
of *Siegfried*. Golther at one point offers Mone's *Einleitung* as
Wagner's source;[28] perhaps the case should rest open.

Both Mone's *Einleitung* and Wilhelm Müller's *Geschichte
und System der altdeutschen Religion* (Göttingen, 1844) were in
the Dresden Royal Library, though it appears Wagner borrowed neither work. Müller's *Altdeutsche Religion* obviously
derives heavily from Grimm, but for Wagner, 'always
hankering after definite, clear and explicit images',[29] it had
certain advantages over the *Deutsche Mythologie*. Where
Grimm inundates his reader with detail Müller simplifies and
selects, and his version produces many more of the clear-cut
images Wagner was seeking. It is, however, particularly in
the instances where Müller departs from Grimm and includes
material from other sources that we seem to detect his
influence on Wagner. We should perhaps keep an open mind
on Müller's *Altdeutsche Religion* too.

Hermann Schneider suggests that Wagner drew on another
of Müller's works, 'Wilhelm Müller's book on the "German
Heroic Sagas" which came out in 1841'.[30] This was the year
in which Wilhelm Müller published his *Versuch einer mythologischen erklärung der Nibelungensage* (Berlin, 1841), presumably the work Schneider is referring to. He is not specific on
how Müller's *Versuch* might have influenced the *Ring*. Müller
is chiefly concerned to demonstrate that Siegfried is the god
Freyr, an idea which certainly does not appear to have been
particularly influential in Wagner's drama. Nor does there
seem to be any external evidence that Wagner knew the
book; he did however know something of the argument, for
Müller returned to the same theme in an article entitled 'Siegfried und Freyr' which appeared in volume 3 of the *Zeitschrift
für deutsches Alterthum* (pp. 43–53). A synopsis of Müller's

---

[28] *Grundlagen*, p. 95.
[29] 'der überall nach bestimmten, deutlich sich ausdrückenden Gestalten verlangte'
(*ML*, p. 273).
[30] 'dem 1841 erschienenen Buch von Wilhelm Müller über die "Deutsche Heldensage"' (Schneider, 'Altertum', p. 109).

theory was also provided in the introduction of Vollmer's edition of the *Nibelungenlied*.[31]

During his Paris years Wagner had connections in the book-trade,[32] and French scholars have been at pains to establish which works of their own literature may have left their imprint on the *Ring*. One such work is Alexandre Dumas's 'Les Adventures merveilleuses du Prince Lyderic', first published in the Paris magazine *Musée des familles* in the autumn of 1841, shortly before Wagner returned to Germany. Dumas's 'Lyderic' is cited by Ernest Newman, whose attention was drawn to the work by an article by Henri Colomb.[33] 'Lyderic' is something of a mid-nineteenth-century equivalent of the eighteenth-century *Volksbuch*. There is no doubting that the prince's adventures are based on those of Siegfried, though it is less certain that Wagner, chancing upon the saga in this disguise, would have recognized as much, particularly since he was at this stage less familiar with the Siegfried saga than he later became. Neither the Sleeping-Beauty-style kiss with which both heroes wake their heroines nor the mere fact that Dumas draws on more than one version of the saga is exclusive evidence of the influence of 'Lyderic' on Wagner's Nibelung drama, and since these are the factors on which Colomb bases his argument his case looks rather thin. If Wagner took anything from Dumas's *oeuvre*, it is to be found in the young Siegfried's love of animals and his preoccupation with the absent mother.

There is, incidentally, nothing to show that Wagner knew the eighteenth-century *Volksbuch* either, although it had recently been republished by Simrock (see Chapter 1).

J. G. Prodhomme, writing in the *Mercure de France*,[34] suggests that Wagner read Edélestand du Méril's *Histoire de la poésie scandinave* (Paris, 1839) with its treatise on Old Norse poetry and translations of selected samples. The strongest point in Prodhomme's argument is that du Méril's *Histoire* was published in Paris by Wagner's in-laws, Brockhaus and

---

[31] *Der Nibelunge Nôt*, pp. xxxvii–xxxix.

[32] e.g. his brother-in-law Avenarius (*ML*, pp. 166 f., 180 f.).

[33] Newman, *Life*, ii. 27 n; Colomb, 'Commentaire', in Marcel Herwegh, *Au soir des dieux* (Paris, 1933), 183–217.

[34] 'Une source française de l'Anneau du Nibelung', *Mercure de France*, NS, 279 (1937), 15 Oct., 439 ff.

Avenarius, the year Wagner himself arrived there. The weak point is that Prodhomme assumes Wagner knew nothing further of the Scandinavian source material prior to meeting Ettmüller in Zurich. There is in fact no inner necessity for Wagner to have known du Méril's work, and perhaps when he first arrived in France, with his French not very secure, Wagner would have been more interested in the German books that his publishing relatives could offer.

Finally, we have to decide whether or not Wagner knew the *Ragnar-Lodbroks-Saga* and *Norna-Gests-Saga*, which appeared as the fifth and final volume of von der Hagen's *Nordische Heldenromane* in 1828. Probably he did not. Deryck Cooke suggests that Wagner may have borrowed this volume along with the *Völsunga saga* from the Royal Library at Dresden, but the Royal Library did not contain the fifth volume of the *Nordische Heldenromane*. There was of course a long gap between the first four volumes of the series in 1814– 15 and the appearance of volume v. When Wagner later wrote to Uhlig from his Swiss exile to ask for the *Völsunga saga* he described the series as the 'Old Norse "chivalric romances" which Hagen—if I'm not mistaken—published in Breslau between 1812 and 1816'.[35] Wagner's detail is in any case not very accurate; but from the approximate dates he gives it looks as though he did not know of the publication of the final volume in the series and almost certainly had not read it.

[35] 'altnordische "Ritterromane" . . . die Hagen—wenn ich nicht irre—1812 bis 1816 in Breslau herausgab' (letter to Theodor Uhlig, 12 Nov. 1851, *Dokumente*, p. 58).

PART II

5

The Work of 1848,
i: *Der Nibelungen-Mythus*

THE first of Wagner's literary works with real bearing on the *Ring*,[1] the *Nibelungen-Mythus* of 1848, contains most of the future tetralogy in embryo. With the exception of Erda and Loge all the *dramatis personae* are present, and almost the whole action is there in outline. Wagner later added items, such as the Rhinemaiden's opening scene, and also left things out, such as Siegfried's revenge expedition, but basically the plot, if not the philosophy, was complete. His dramatic concept of the Nibelungs was already formed, therefore, and our initial enquiry is to establish what it was formed from, whereby the availability of primary source material is perhaps the dominant consideration.

Let us remind ourselves of Wagner's bibliographical progress at this point. When he completed the *Mythus* on 4 October 1848 a number of items from his ultimate reading-list, some important, were still missing. Of the non-primary sources he had not yet seen the third volume of Simrock's *Amelungenlied*, published in 1849, nor the issues of the *Zeitschrift für deutsches Alterthum* of late 1848 and early 1849, including Weinhold's comprehensive study of Loge. Among the primary sources Wagner had of course not yet read Simrock's *Edda*, first published in 1851. In consequence there were a number of 'Götterlieder' probably still unknown to him: *Grógaldr*, *Hyndluliódh*, *Rígsmál*, and three which were later to influence his *Ring* poem: *Fiölsvinnsmál*, *Oegisdrecka*, and *Hrafnagaldr Odhins*. The *Oegisdrecka*, or *Locasenna* as it is

---

[1] *Die Wibelungen* is now thought to have been written after the *Mythus*: see Deathridge, Geck, and Voss, *Wagner Werk-Verziechnis*, p. 329. Although on related themes, it has in any case no structural parallel with the *Ring* drama and only a marginal overlap in material.

otherwise known, is a Loge poem, and Wagner's ignorance of both this and Weinhold's article on the demigod may account for the absence of Loge from the *Mythus* of 1848. Also missing from Wagner's bibliography at that date was Ettmüller's Edda translation, and with it all the later part of the 'Heldenlieder' cycle, dealing with developments after Siegfried's death. In the absence of both Simrock and Ettmüller Wagner was dependent for the Eddic heroic poems exclusively on the Grimm brothers' translation, which stopped short at *Helreidh Brynhildar*. Wagner knew the continuation of the saga in the *Snorra Edda*, in the *Nibelungenlied* and probably in the *Thidreks saga* too; yet with two of his primary sources: *Das Lied vom Hürnen Seyfrid* and now the *Poetic Edda*—appearing to conclude matters with Siegfried's death, Wagner must have found at least reinforcement for his decision to do the same.

Another 'absentee' at this stage was the *Völsunga saga*. It perhaps seems strange that Wagner should have embarked on his Nibelung project without having read a work which has generally been considered essential for the drama. Deryck Cooke, for instance, writes: '*Völsunga saga* was really Wagner's main source, in the sense that it alone offered him a continuous and coherent tying-together of most of the main episodes he used for the plot of *The Ring*',[2] while Hermann Schneider singles out the *Nibelungen-Mythus* as showing unambiguously the influence of the *Völsunga saga*: 'No doubt about it, in seeking to immerse himself in the Nibelung saga the Dresden Kapellmeister derived ideas and impetus above all from the prose History of the Welsungs (*Völsunga saga*), which had already been translated into German. The sketch of 1848 [*Nibelungen-Mythus*] reveals this all along the line.'[3] Only Ashton Ellis correctly surmises that Wagner did not know the *Völsunga saga* in the period when he wrote the *Mythus* (see Chapter 3).

Wagner was of course himself well aware of the central

[2] *World*, p. 113.
[3] 'Kein Zweifel, der Dresdener Kapellmeister, der sich in die Nibelungensage zu versenken suchte, schöpfte Begriffe und Anregungen vor allem aus der prosaischen Geschichte der Welsungen (Völsungasaga), die bereits ins Deutsche übersetzt war. Der Entwurf von 1848 verrät das auf Schritt und Tritt.' ('Altertum', p. 113.)

importance of the *Völsunga saga* to his theme even before he had read it, and as his correspondence with Uhlig shows, he had made 'every conceivable effort' to obtain it (see Chapter 3). However great the influence of the *Völsunga saga* was on the finished *Ring*, though, the *Mythus* was written without it. Some of what is usually atttributed to the *Völsunga saga* in Wagner's Nibelung drama came to him second-hand, via the knowledge of his contemporaries; the larger part was derived direct from the Eddas.

Not the *Völsunga saga* then, but the Eddas provided most of the material for the *Mythus*. It was in the summer of 1848 that Wagner began his raids on the Royal Library Edda holdings; by early October he had read, or reread, the 'Heldenlieder' of Grimm and the 'Götterlieder' translations by Studach and Legis. Given this surge of interest in the Edda, which lasted until the close of the year and subsequently took in Ettmüller's translation of the Nibelung poems as well, we can imagine that Wagner had a second look at those Edda translations on his own shelves: Ettmüller's *Vaulu-Spá*, the *Mythologische Dichtungen* of Majer, and the Rühs *Snorra Edda*. Once again the evidence forces us to disagree with Hermann Schneider, who suggests that only the *Snorra Edda* was influential in the *Mythus*, and that the *Poetic Edda* had to wait until Wagner met Ettmüller in Zurich.[4]

At a rough estimate, around three-quarters of the primary source derivation in the *Mythus* is from the Eddas. Of the various beings, human, divine, and other, who inhabit the world of the *Ring*, only the Rhinemaidens have no Eddic connections. Of the plot, the Eddas are the sole source for the two episodes which initiate the gods' dilemma: the building of the fortress by giant workmen from the *Snorra Edda* and the theft of the dwarf's gold to pay the gods' obligation, again from Snorri and *Sigurdharkvidha Fafnisbana önnur*. Wagner's combination of the two separate stories was one of his most brilliant 'engineering' feats,[5] surpassed perhaps only by another fusion of separate events: he took the story of

---

[4] 'Altertum', pp. 113, 115.
[5] For further details see e.g. Golther, *Grundlagen*, pp. 30–3 and Cooke, *World*, pp. 176–81.

Brünnhilde's disobedience from the fight between two unknowns, Agnar and Hialmgunnar, and transferred it to the account of Siegmund's death.[6] While Siegmund's death derives ultimately from the *Völsunga saga*, by a route we shall later trace, the Agnar–Hialmgunnar fight and Brünnhilde's punishment are told in the Eddic poem *Helreidh Brynhildar*.[7]

On the next section of the *Mythus*, dealing with Siegmund's history and Siegfried's earlier forebears, the Eddas are silent. However from now on, with the exception of Siegfried's death, that Eddas have at least shared responsibility for all further primary source influence on Wagner's Nibelung concept. They contribute to Siegfried's upbringing under Mime's tutelage and the forging of his sword, to the dragon fight, the bird-song, Mime's death, and the waking of Brünnhilde. They influence his arrival at Gunther's court and his betrothal to Gutrune, the winning of Brünnhilde for Gunther and her later accusations against him, and finally her self-immolation on Siegfried's funeral pyre.[8]

For Siegfried's death itself Wagner turned to the *Nibelungenlied*. Some of the preliminaries already show the influence of the German epic: the oath Siegfried swears to clear himself of Brünnhilde's charges[9] and the betrayal of Siegfried's vulnerability.[10] From the hunting scene of what is now Act III of *Götterdämmerung* through to the bearing-back of Siegfried's body, the *Nibelungenlied* which Wagner affected to discount as a source for his drama (see Chapter 1) reigns supreme—and quite rightly, for its account of Siegfried's last hours is one of the highlights of the poem. From the *Nibelungenlied* Wagner took Siegfried's ebullient mood and the spear thrust between his shoulder-blades while he is off guard, the abortive attempt to avenge himself on Hagen with his shield, and the

---

[6] Again, Golther gives an excellent account, *Grundlagen*, pp. 51–4; Cooke also mentions the episode, *World*, p. 114.

[7] Brünnhilde's disobedience is also good *Vs* material, and it is in fact at least equally likely that Wagner took the episode from the purveyors of *Vs* subject-matter mentioned in this ch.

[8] Told both in the Eddic poems from *Sigurdharkvidha Fafnisbana fyrsta* through to *Sigurdharkvidha Fafnisbana thridhja* and in the *Snorra Edda* (*SnE*).

[9] *NL*, Aventiure 14.

[10] *NL*, Aventiure 15.

final thought for his wife as he lies dying on the grass.[11]

The balance of evidence also favours the *Nibelungenlied* as the source of Wagner's Rhinemaidens. In the *Mythus* they appear to Siegfried on the banks of the Rhine just before his death, and again to receive the ring back from Brünnhilde, much as in the present final act of *Götterdämmerung*. Alberich's theft of the gold from the Rhinemaidens, the scene with which the *Ring* now opens, was not invented until Wagner came to work on *Das Rheingold*.

At first sight the *Nibelungenlied* may not seem a very promising source for the Rhinemaidens, since it does not contain any; in fact, as Jacob Grimm concluded in the chapter 'Elemente' of his *Mythologie*, Rhinemaidens are not found anywhere in German tradition.[12] Even less does the *Nibelungenlied* or any other source connect water-nymphs with the gold in the Rhine, though it was well known that the gold was there,[13] not least because that is where it is put in some of the Nibelung literature. In the *Nibelungenlied* itself, at Aventiure 19, Hagen sinks the hoard in the Rhine after Siegfried's death; but in the poem no water-maidens rise from the waves to receive it.[14]

The closing stage business of Wagner's *Mythus* and later of his drama, therefore, did not derive from the Middle High German epic. For the earlier Rhinemaiden scene, on the other hand—the encounter with Siegfried on the banks of the Rhine—the *Nibelungenlied* did deliver what can be viewed as a prototype. In Aventiure 25 Gunther and his kinsmen, some years after Siegfried's death, are journeying to the fatal feast at Etzel's court from which none of them will return alive. They come to the Danube, and Hagen, looking for a crossing, finds some mermaids bathing in a pool: it is not clear how many, but there are at least two. Hagen takes their clothes away and in the ensuing conversation the mermaids prophesy the death of Hagen and all his party.

It has generally been agreed that this episode from the

---

[11] *NL*, Aventiure 16.
[12] p. 567.
[13] Ibid.
[14] In *Das Lied vom Hürnen Seyfrid* (*HS*) it is Siegfried himself who dumps the hoard in the Rhine.

Nibelungenlied served as Wagner's stimulus.[15] However, an alternative source for this scene has recently been suggested by Nancy Benvenga. Saxo Grammaticus relates an incident concerning Hother, here a rival of Baldur: while out hunting Hother goes astray in the mist and comes across some wood-maidens in a hut; they advise Hother not to attack Baldur, the demigod, with weapons and then vanish into the mist. We have suggested that Wagner did not read Saxo (see Chapter 2); but as Dr Benvenga points out, the episode is retold in Frauer's book on valkyries, which was in Wagner's library.[16]

Weighing up the two postulated sources for the scene—the *Nibelungenlied* and Saxo—it is soon apparent that the evidence for Saxo falls short of that for the *Nibelungenlied*. The wood-maidens' advice on how not to kill his enemy is hardly as relevant to Wagner's riverside scene as is the warning of impending death issued to Hagen on the banks of the Danube. Two other factors cited by Dr Benvenga in support of her argument do not really advance the case. If the wood-maidens of Saxo, like Wagner's Rhinemaidens, address their hero by name,[17] so too do the mermaids of the *Nibelungenlied*. Secondly, Dr Benvenga writes: 'Even the original name Wagner gave the Rhinemaidens—*Wasserjung-frauen* [water-maidens] echoes Frauer's *Waldjungfrauen* [wood-maidens]'.[18] The 'Rheintöchter' (Rhine-daughters) of the *Ring* do indeed appear as 'Wasserjungfrauen' in *Siegfrieds Tod*. Yet this was not their original name; they start out life in the *Mythus* as 'Meerfrauen' ('sea-ladies'), which by the end of the sketch has become 'Wasserfrauen' ('water-ladies').[19] The term the *Nibelungenlied* uses, which like most of these can be loosely translated as 'mermaids', is 'Meerweiber' ('sea-women'). The progression to 'water-maidens' was evidently not from Saxo's 'wood-maidens' but from the 'sea-women' of the *Nibelungenlied*, via the 'sea-ladies' and 'water-ladies' of the *Mythus*.

This is not to say that the Saxo episode had no influence on

---

[15] e.g. by Golther, *Grundlagen*, pp. 34–5; Cooke, *World*, p. 139.
[16] *Kingdom on the Rhine* (Harwich, 1983), 37.
[17] Ibid.
[18] Ibid.
[19] *Skizzen*, pp. 30, 33.

Wagner's drama; one further parallel cited by Nancy Benvenga, that both Hother and Siegfried are out hunting at the time of their encounter, could on the contrary have been quite important. It may have been the hunting context of the Hother incident which inspired Wagner to introduce a comparable scene for Siegfried into the final hunt of the *Nibelungenlied* death-story.

That it was definitely 'water-game'[20] Wagner had in mind for this scene and not their sylvan counterparts is clear from his first description of the prototype Rhinemaidens in the *Mythus*: 'three mermaids with swans' wings'.[21] Wagner's first thought was obviously of swan-maidens, which he knew about in detail from Jacob Grimm's treatment of the subject in the chapter 'Weise Frauen' of his *Mythologie*. Grimm describes how these fly through the air as swans but become transformed into women when they touch down to bathe by taking off their feather dresses or wings. Whoever removes these gains power over them.[22] Hagen knew this, and Grimm writes that the prophesying mermaids of the *Nibelungenlied* come into the swan-maiden category: 'although it is not expressly stated, the three [*sic*] prophesying mermaids whose clothing Hagen had removed are precisely this'.[23] He quotes the well-known line:

> sie swebten *sam die vogele* vor ihm ûf der fluot.[24]
> They hovered just like birds before him on the flood.

Aventiure 25 of the *Nibelungenlied* had yielded mermaids on the banks of the Danube. Wagner wanted them in the Rhine, traditionally the seat of Gunther's court, scene of Siegfried's last hunt, and final home of the gold. One primary source had already half-achieved such a transfer. The *Thidreks saga*, which we have suggested Wagner already knew at this point (Chapter 3), tells what is essentially the

---

[20] 'Wasserwild'—Siegfried's ungallant term for the Rhinemaidens when he rejoins his hunt companions (*Skizzen*, p. 31).
[21] 'Drei Meerfrauen mit Schwanenflügel' (*Skizzen*, p. 30).
[22] Grimm, *Mythologie*, pp. 398–9.
[23] 'obgleich es nicht ausdrücklich gesagt wird, die drei [*sic*] weissagenden meerweiber, denen Hagne das gewand weggenommen hatte, sind eben solche' (*Mythologie*, p. 399).
[24] Ibid.

same story as Aventiure 25 of the *Nibelungenlied*, though with a distinctly unfortunate outcome for the prophesying maidens. Only the geography is different, and, we must admit, somewhat insecure, for in the *Thidreks saga* the scene is set at the confluence of the Danube and the Rhine.[25] As for the mermaids, who are bathing in a pool nearby, the *Thidreks saga* says: 'these mermaids had gone into this pool from the Rhine'.[26]

According to the *Thidreks saga* then, the mermaids, while found by the Danube, originated in the Rhine. Grimm evidently left the *Thidreks saga* out of account when he wrote on the absence of spirits in the Rhine (see above); he was, of course, only speaking of the native tradition, and since the *Thidreks saga*, while based on German sources, was written in Scandinavia, Grimm's omission is both understandable and valid.

Das Nibelungenlied, then, *Thidreks saga*, perhaps Saxo: these inspired the first of the two Rhinemaiden scenes in Wagner's *Mythus*. But what of the second, that closing scene of Wagner's dramatic concept, where the daughters of the Rhine rise in the water for their ring? We revert to the *Nibelungenlied*, only this time not to the text but to some of the illustrations. We suggested earlier that Wagner's main reason for buying the Pfizer edition of the *Nibelungenlied* was its inclusion of the woodcuts by Julius Schnorr von Carolsfeld and Eugen Neureuther. Their attitude to textual authority varied. Sometimes their illustrations are in strict accordance with the text; one such woodcut to Aventiure 25 shows Hagen and the mermaids. Elsewhere they exercise great licence in their treatment of the text, and in an illustration to Aventiure 19 we discover the mermaids again, circling expectantly in the water while Hagen pours the hoard into the Rhine.[27]

The waiting water-nymphs were a piece of pure Romantic invention. To Wagner the scene suggested both the closing stage effect of his drama and, more generally, the bond between Rhinemaidens and gold which was later to span the

[25] *Thidreks saga* (*Thss*), ii. 340.
[26] 'Diese Meerweiber waren aus dem Rhein in dieses Wasser gegangen' (ii. 342).
[27] See frontispiece. The illus. to Aventiure 25 is on p. 281, that to Aventiure 19 on p. 211 of Pfizer's *Der Nibelungen Noth*.

whole *Ring*. In sum, we concede that Wagner's Rhinemaidens owe their existence primarily to the *Nibelungenlied*, whereby the debt is in part to the poem and in great measure to the particular contemporary visual interpretation by Schnorr von Carolsfeld and Neureuther.

After eliminating influence from the Eddas and *Nibelungenlied* and disregarding for the moment that of the *Thidreks saga* and *Das Lied vom Hürnen Seyfrid*, we are left with certain areas of the *Mythus* which defy derivation from any primary source other than the *Völsunga saga*: Siegfried's ancestors, the history of the sword, and the magic potion by which Siegfried loses all memory of Brünnhilde. Since the *Völsunga saga* itself still eluded Wagner the material must have come to him via the efforts of others, and the question now facing us is the direction from which his help came.

Wagner could without much difficulty have pieced together an overall outline of the *Völsunga saga* plot from the introductions to other primary sources: Vollmer's introduction to the *Nibelungenlied*, for example, is quite good on the later part of the *Völsunga saga*,[28] while for the beginning von der Hagen provided an adequate synopsis in the introduction to his 1812 *Edda*.[29] Such accounts could have enlightened Wagner; he would not, however, have found them particularly inspiring. For this other, more particular accounts were to hand.

The magic potion came to him, with little doubt, from Fouqué, whose dramatic poem *Sigurd der Schlangentödter*[30] is so staunchly based on the *Völsunga saga*. We shall have cause to refer to Fouqué's *Sigurd* so often in our study of Wagner's two Siegfried dramas that it is probably safe to say that from the point where his dramatic poem opens in Reigen's smithy Fouqué is chiefly responsible for all *Völsunga saga* material in Wagner's *Mythus*.

The early part of the *Völsunga saga*, however, is not

---

[28] *Der Nibelungen Nôt*, pp. vii ff.

[29] pp. xxviii ff.

[30] Fouqué, *Ausgewählte Werke*, vol. i (Halle, 1841). The administering of the potion is given on pp. 104–7. Fouqué was probably also Wagner's source for the history of Brünnhilde's disobedience, which appears on pp. 62–3 of *Sigurd*; see n. 7 above.

covered by Fouqué's poem, and Wagner must have turned elsewhere for Siegfried's ancestors and the history of the sword. They are in any case more complex issues. The *Mythus* runs along slightly different lines from the present *Ring* at this point, particularly where the internal motivation is concerned, and we perhaps need to remind ourselves of Wagner's original conception:

At last this hero is due to be born, into the Wälsung family. Wodan has fertilized a barren marriage of the Wälsungs with one of Holda's apples which he gave the couple to eat; twins, Sigemund and Sigelind (brother and sister), spring from the marriage. Sigemund takes a wife, Sigelind weds a husband (Hunding), but both their marriages remain childless. In order to beget a true Wälsung brother and sister now pair with each other. Hunding, Sigelind's husband, learns of the offence, repudiates his wife, and attacks Sigmund.[31]

There follows the famous account of Brünnhilde's disobedience, and with it the first mention of the sword:

The valkyrie Brunhild defends Sigmund against the orders of Wodan, who has decreed Sigmund's fall in expiation of the offence. Sigmund, shielded by Brünhild, is just brandishing the sword, which Wodan himself once gave him, to deliver Hundling's death-blow when the god intercepts the blow with his spear and the sword breaks in two against it. Sigmund falls.[32]

After Brünnhilde's punishment the story resumes with Sieglinde:

The rejected Sigelind gives birth in the wilderness to Siegfried (who is to bring peace [Frieden] through victory [Sieg]). Reigin

[31] 'Im Geschlecht der Wälsungen soll endlich dieser Held geboren werden: eine unfruchtbar gebliebene Ehe dieses Geschlechtes befruchtete Wodan durch einen Apfel Holda's, den er das Ehepaar geniessen liess: ein Zwillingspaar, Sigemund u. Sigelind (Bruder u. Schwester) entspringen der Ehe. Sigemund nimmt ein Weib, Sigelind vermählt sich einem Manne (Hunding); ihrer beiden Ehen bleiben aber unfruchtbar: um einen ächten Wälsung zu erzeugen begatten sich nun Bruder u. Schwester selbst. Hunding, Sigelind's Gemal, erfährt das Verbrechen, verstösst sein Weib u. überfällt Sigmund mit Streit.' (*Skizzen*, p. 27.)

[32] 'Brunhild, die Walküre, schützt Sigmund gegen Wodan's Geheiss, welcher dem Verbrechen zur Sühne ihm den Untergang beschieden hat; schon zückt unter Brünhilds Schild Sigmund zu dem tödlichen Streiche auf Hunding das Schwert, welches Wodan ihm einst selbst geschenkt, als der Gott den Streich mit seinem Speer auffängt, woran das Schwert in zwei Stücken zerbricht: Sigmund fällt.' (*Skizzen*, pp. 27–8.)

(Mime), Alberich's brother, went to Sigelind's assistance from out of the rocky abyss as she cried out in labour. After giving birth she dies, having told Reigin her history and entrusted the boy to him. Reigin brings Siegfried up, teaches him the blacksmith's trade, informs him of his father's death, and obtains for him the two pieces of Sigmund's broken sword, out of which Siegfried forges the sword Balmung under Mime's direction.[33]

Various features stand out in the Wälsung history of Wagner's *Mythus.* For a start, it is brief. Secondly, compared with the complexity of the *Völsunga saga* it is drastically cut. Wagner says nothing of the twins' ancestors, and he has left out their nine brothers. Gone, too, are Siegmund's nephews and at least one of his marriages with its offspring.[34] Thirdly, what remains of the Wälsung history has been radically telescoped so that the *Völsunga saga* account of Signe, Sinfiötli, and Siggeir is totally fused with that of Hiordis, Siegfried, and the Hundings. Signe, Siegmund's twin, in effect incorporates Hiordis, Siegfried's mother;[35] the son of the incestuous union of the twins, Sinfiötli in the *Völsunga saga,* has become Siegfried; Siggeir, Signe's husband and Siegmund's enemy of the saga, is now named Hunding, head of the rival house which in the *Völsunga saga* kills Siegmund in a quarrel over Hiordis. The effect of Wagner's compression is to leave us with the barest bones of the story: the twins, their union, Siegfried's conception, and Siegmund's death in consequence—embellished with one or two stray motifs from the saga: the barren marriages, the apple of fertility, the long pregnancy.

It is *Völsunga saga* material, but if we are looking for the intermediary through whom it arrived at Wagner we can

---

[33] 'Die verstossene Sigelind gebiert in der Wildnis nach langer Schwangerschaft Siegfried (der durch Sieg Friede bringen soll): Reigin (Mime) Alberichs Bruder, ist, als Sigelind in den Wehen schrie, aus Klüften zu ihr getreten u. hat ihr geholfen: nach der Geburt stirbt sie, nachdem sie Reigin ihr Schicksal gemeldet u. den Knaben diesem übergeben hat. Reigin erzieht Siegfried, lehrt ihn schmieden, meldet ihm den Tod seines Vater's und verschafft ihm die beiden Stücken von dessen zerschlagenem Schwerte, aus welchen Siegfried unter Mime's Anleitung das Schwert Balmung schmiedet.' (*Skizzen,* p. 28.)

[34] In *Vs* Siegmund marries twice (Borghild, Hiordis) and has a son from each marriage (Helgi, Siegfried), in addition to the incestuous union with his twin sister which results in Sinfiötli. The barren marriage was Wagner's own idea and was later dropped.

[35] Her name, Sieglinde, is taken from *NL* and *HS.*

say with safety that a highly detailed and near-nigh ver-
batim account, such as Simrock later included in his
Amelungenlied,[36] would not have been most to his purpose. In
any case, the first edition of Simrock's *Amelungenlied*, the one
available to Wagner, did not yet include the Wälsung history;
perhaps Simrock was finding the *Völsunga saga* as difficult to
locate as Wagner was. A suitably taut and schematic account
was, however, available, in Wilhelm Grimm's *Deutsche
Heldensage*. In the essay 'Ursprung und Fortbildung' at the
end of his *Heldensage* Grimm includes a brief section, based
on the *Völsunga saga*, headed 'Siegfried's Ancestors':

Völsung, Sige's grandson and Sigurd's grandfather, is conceived
after Odin has sent a fertility-promoting apple to his father Rerir.
His mother does not give birth to him, but at six years old he is cut
from her body. Völsung's most famous son, Siegmund, takes in his
sister Signe without recognizing her; their son is Sinfiötle. After
leading a wild, animal existence for a while, transformed into
wolves, the two of them, father and son, avenge Völsung's death
on Siggeir, Signe's husband. Sinfiötle is poisoned by his step-
mother Borghild by means of a drink; Siegmund then marries
Hiordis and she gives birth to Sigurd, but not until after Siegmund's death.[37]

This is still a long way from Wagner's version. Wilhelm
Grimm was not at liberty to make far-reaching amalga-
mations of plot and character as the poet might; he could
only select and omit. Nevertheless, his Wälsung history
shows striking parallels with Wagner's. One notices immediately that Siegmund's brothers, nephews, and son Helgi are
out, the fertility-apple and long pregnancy are in. While
Grimm goes back a generation further than Wagner, starting

---

[36] Later edns. of the *Amelungenlied* incorporated it into Wieland's narrative in
Abenteuer 16–21 of the section 'Wieland der Schmied'.

[37] '*Siegfrieds Ahnen*: Völsung, Siges Enkel, Sigurds Grossvater, wird erzeugt,
nachdem Odin seinem Vater Rerir einen fruchtbringenden Apfel gesendet hat. Er
wird nicht geboren, sondern, schon sechs Jahre alt, aus Mutterleib geschnitten.
Völsungs berühmtester Sohn, Siegmund, nimmt, ohne sie zu kennen, seine Schwester Signe bei sich auf; ihr Sohn ist Sinfiötle. Beide, Vater und Sohn, nachdem sie eine
Zeit lang, in Wölfe verwandelt, ein wildes thierisches Leben geführt, rächen an
Siggeir, dem Gemahl der Signe, den Tod Völsungs. Sinfiötle wird von seiner
Stiefmutter Borghild durch einen Trank vergiftet; Siegmund vermählt sich hierauf
mit Hiordis, und diese gebiert, doch erst nach seinem Tode, den Sigurd.' p. 337.

with the conception of the twins' father Völsung, his interest in the preceding generations is minimal. Above all, Grimm makes no mention of the episode at Signe's wedding where the divine guest thrusts into the tree a sword which only Siegmund can claim; nor, although he mentions Völsung's death, does he say what set in motion the enmity between Siggeir and the Wälsungs.

With the suggestion that it was Wilhelm Grimm's account in the *Heldensage* that inspired the Wälsung history of Wagner's *Mythus*, we turn now to the last indisputably *Völsunga saga*-derived area of the *Mythus*: Siegmund's sword. The first thing that strikes us is that both Wagner and Grimm have omitted the distinctive episode of the sword in the tree from their respective Wälsung histories. The first we hear of a sword in the *Mythus* is at Siegmund's death, where he is wielding the sword 'which Wodan himself once gave him' (see above). Wagner evidently knew the story of the god's gift, but, like Grimm,[38] kept it separate from the history of Siegfried's forebears. Later, after reading the *Völsunga saga* for himself, Wagner was able to settle on a way of introducing Wotan's sword-gift into the *Ring*,[39] but jealousy over the sword, which in the *Völsunga saga* causes the rift between Siegmund and his brother-in-law, attains only a minor or negligible role in Wagner's Nibelung drama.[40]

The *Thidreks saga* may have helped Wagner with his Rhinemaidens; more evidence of its influence can be seen in the story of Siegfried's birth and childhood (see above). The repudiation of Sieglinde in the *Mythus* and her *accouchement* in the wilderness are obvious *Thidreks saga* contributions. The *Thidreks saga* as much as the Eddas might be responsible for Siegfried's upbringing in the smithy, while for his appren-

---

[38] *Heldensage*, p. 381.

[39] In his sketches for *Die Walküre* Wagner had opted for a different handling of the episode, with Wotan entering while Siegmund was present and placing the sword in the tree then and there (*Skizzen*, pp. 204, 211–12, 235).

[40] In the prose plan to *Die Walküre*, where Siegmund withdraws the sword in Hunding's presence, Wagner included jealousy over the sword as one of the factors in Hunding's enmity towards Siegmund; even here, however, the sword does not start their dispute, which stems from the past (*Skizzen*, p.235). The jealousy over the sword was dropped when Wagner decided to omit Wotan's stage entry in Act I.

ticeship to Mime the honours must be divided between the *Thidreks saga* and *Das Lied vom Hürnen Seyfrid*.

The chief contribution of the last of Wagner's major primary sources, *Das Lied vom Hürnen Seyfrid*, was undoubtedly to the giants and dwarfs of Wagner's drama, and in particular to the state of tension between the two communities. According to this source, the dwarfs live in constant fear under the oppressive tyranny of the giant Kuperan. After Kuperan's death their king, Eugel, tells Siegfried:

> Kuperan der falsche Riese bezwang hier unsern Berg,
> Darin wohl tausend Zwerge ihm wurden unterthan:
> Wir zinsten unser Eigen dem ungetreuen Mann.
> Nun habt ihr uns erlöset, wir wurden alle frei.[41]
>
> Kuperan the wicked giant seized our mountain home;
> A thousand dwarfs or more became his subjects then,
> Our freehold turned to lien by that treacherous man.
>
> But you have now delivered us, and we have all been freed.

In some kind of unholy alliance with Kuperan is the dragon, whom the dwarfs also fear, though the poem never clearly explains the relationship between the two.[42] Some confusion also arises concerning the hoard, which Siegfried pockets after he has dispatched Kuperan and the dragon, under the impression that it belonged to one or other of these. This time the poem itself is quite clear that the hoard belongs to the dwarfs, who are keeping it in the dragon's rock. Eugel complains to Siegfried that the giant has taken over their property (see above), though it must be admitted that the dwarfs seem quite free in the poem to come and go with their treasure as they please.

From this material Wagner fashioned a case history of social conflict generated by the gold. Numerous attempts have been made to interpret Wagner's Nibelung drama strictly in terms of nineteenth-century society, of which Bernard Shaw's is the wittiest.[43] All fall short ultimately, for the

---

[41] Simrock, *Heldenbuch*, iii. 203.

[42] Perhaps simply landlord and tenant, since Kuperan keeps the key to the dragon's rock (Simrock, *Heldenbuch*, iii. 181).

[43] *The Perfect Wagnerite* (London, 1898).

reason that Wagner was too true to his sources and his own broader dramatic instincts: his gods are too much gods, his giants giants, his Nibelungs dwarfs, to double successfully as schematic kings and aristocrats, gentry and proletariat, capitalists, plutocrats, or whatever. Nevertheless, there is no doubt that certain sections of the *Mythus* were intended, if not as an exact analogy of contemporary society, then at least as a symbolic demonstration of the baleful workings of capitalism.

In the *Mythus* Wagner relates that as Alberich gains power over his fellow Nibelungs through the ring—'The giant race . . . is disturbed in its state of rough-hewn complacency . . . They view with apprehension the Nibelungs forging marvellous weapons which one day, in the hands of human heroes, will bring about the downfall of the giants.'[44] And so the giants demand Alberich's hoard in return for building the gods' fortress, and while they make little use of their booty they retain the Nibelungs in servitude:

Now the giants arrange for the hoard and the ring to be guarded on Gnitaheide by a monstrous dragon. Through the ring the Nibelungs, along with Alberich, remain in servitude . . . Wretched and malicious, the Nibelungs languish on in unproductive activity . . . The soul, the freedom of the Nibelungs lies useless, guarded under the belly of the idle dragon.[45]

In the *Mythus* the plight of the Nibelungs is the chief item in the gods' guilt, for they had it in their power to restore freedom to the dwarfs but chose not to do so: 'Out of the depths of Nibelheim the consciousness of their guilt rumbles up to them; for the bondage of the Nibelungs is not broken, control has merely been snatched from Alberich . . .'[46]

[44] 'Das Geschlecht der Riesen . . . wird in seinem wilden Behagen gestört . . . sie sehen mit Sorge die Nibelungen wunderbare Waffen schmieden, die in den Händen menschlicher Helden einst den Riesen den Untergang bereiten sollen.' (*Skizzen*, p. 26.)

[45] 'Nun lassen die Riesen den Hort u. den Ring auf der Gnitaheide von einem ungeheuren Wurm hüten: durch den Ring bleiben die Nibelungen mit Alberich zugleich in Knechtschaft . . . elend u. tückisch schmachten die Nibelungen in fruchtloser Regsamkeit fort . . . unter dem Bauche des müssigen Wurmes liegt nutzlos die Seele, die Freiheit der Nibelungen gehütet.' (*Skizzen*, pp. 26–7.)

[46] 'Aus den Tiefen Nibelheim's grollt ihnen das Bewusstsein ihrer Schuld entgegen; denn die Knechtschaft der Nibelungen ist nicht gebrochen, die Herrschaft ist nur Alberich geraubt . . .' (*Skizzen*, p. 27).

Redemption in the *Mythus* takes the form of breaking the
Nibelungs' bondage, as Brünnhilde announces on returning
the ring to the Rhine: 'Let the Nibelungs' bondage be
broken, the ring shall bind them no longer . . .'[47] Later
developments in the *Ring* rather eclipsed the central Nibelung
theme of the *Mythus*. Rival attractions[48] claimed attention,
and in the final cataclysm of *Götterdämmerung* the Nibelungs'
freedom is forgotten. Nevertheless, in Wagner's original
Nibelung concept the fate of the oppressed dwarfs was a
major concern.

On to the background of fear and mistrust from *Das Lied
vom Hürnen Seyfrid*, Wagner grafted the story, told in the
Eddas and *Völsunga saga*, of Andvari's gold, robbed by the
gods to pay Hreidmar and his sons Fafner and Reigin the
weregeld for another son, Otur, whom they have
inadvertently slain. The Scandinavian story then tells how
Fafner kills his father and drives away his brother in order to
enjoy sole possession of the gold, which he guards in the
form of a dragon.

Of this, only the middle section—the robbing of the dwarf
to pay the gods' debt—found its way into the *Mythus*. Instead
of the killing of Otur Wagner used that other Edda story, of
the giant workman and the building of the fortress, to
explain the nature of the gods' debt. At the other end of the
story, there is no mention of the parricidal hoard-owner-
cum-dragon Fafner in the *Mythus*.

Even so, combining the story of Andvari's gold with the
account of oppressed dwarfs from *Das Lied vom Hürnen Seyfrid* was not an obvious move. The Scandinavian story had
produced a hoard-owning family with barely a hint that they
were giants;[49] *Das Lied vom Hürnen Seyfrid* features a dwarf-
oppressing giant who, we have just seen, had no hoard. For

---

[47] 'gelöst sei der Nibelungen Knechtschaft, der Ring soll sie nicht mehr binden
. . .' (*Skizzen*, p. 33).

[48] e.g. Freia.

[49] In *Fafnismál* both Fafner and Reigin are called 'Joten' ('giants': Simrock, *Edda*,
pp. 164, 165); but since it is also said in *Sigurdharkvidha Fafnisbana önnur* that Reigin is
'ein Zwerg von Wuchs' ('of dwarf stature': Simrock, *Edda*, p. 156), the ref. to the
brothers as giants is not to be taken too seriously. In the context 'Joten' is probably
no more than a term of abuse.

the polar society of his *Mythus* Wagner created hoard-owning, Nibelung-dominating giants.

Wagner was spared at least some of the imaginative effort
involved, for the task had already been successfully tackled,
albeit along slightly different lines, by Raupach, whose
Nibelungen-Hort was the one Nibelung drama appearing prior
to his own with which Wagner half-owned acquaintance
(see Chapter 4). The prologue of Raupach's drama, based
on a combination of *Das Lied vom Hürnen Seyfrid* and
Scandinavian material, tells of dwarfs and their giant oppres-
sors. Raupach chose to retain the Scandinavian personnel,
and so instead of the 'Giant Kuperan' and the unrelated dragon he gives us the 'Giant [*sic*] Hreidmar', Fafner, and
Reigen.

The part of the hoard story which interested Raupach was
not its central section but the ending, a cautionary tale of
cupidity which deftly illustrates the evil genius of the hoard.
He omits earlier events—Otur's death and Andvari's gold—
and shows instead Hredimar and his sons acquiring wealth
by usurping the Nibelungs' hoard. In a passage which has
more than passing similarity to the account Mime later gives
in *Das Rheingold* of Alberich's regime, the dwarf Eugel describes to Siegfried the dwarf's plight, and then the fate of
Hreidmar and sons:

> Wir wohnen—Nibelungen nennt man uns—
> von Anfang her in diesen Felsenkammern,
> Und uns're Lust was stets, was irgend glänzt,
> Erz oder Stein zu holen aus der Nacht,
> Und manch ein künstlich Werk daraus zu bilden.
> So ward der Hort gesammelt. Dies nun wissend,
> Kam übers Meer der Riese Hreidmar her,
> Und machte sich zum Herren uns'rer Schätze
> Und uns zu Knechten.—Dienen mussten wir,
> In harter Frohne thun, was vormals wir
> Aus freier Lust gethan, bei Tag und Nacht
> Oft schwer gezüchtigt, ihm den Schatz vermehren.
> Die grossen Götter gaben ihm den Lohn:
> Denn seine Söhne—Fafner hiess der eine,
> Der and're Reigen—nach dem Reichtum geizend,
> Erschlugen ihren Vater, da er schlief.
> Als sie nun theilen sollten, stellte Fafner,

Weil er den Schatz allein besitzen wollte,
Mit arger List dem schwächern Bruder nach;
Doch der entfloh, und ward nicht mehr gesehen.
Die grossen Götter straften Fafners Frevel,
Er ward zum Drachen, den Du heut' erschlugst.[50]

We have lived—they call us Nibelungs—
Since time began within these rocky chambers,
And our delight was always, from the night
To fetch whatever shone, ore or gem,
And fashion from it skilful artefacts.
Thus the hoard was gathered. Giant Hreidmar,
Knowing this, now came from over sea
And made himself the master of our treasure
And us his slaves. We were obliged to serve,
In forced labour do, what once we did
Of our free inclination: often punished
Day and night, increase the hoard for him.
The great gods gave him his reward:
For his sons—one by name of Fafner,
The other Reigen—coveting his wealth,
Slew their father while he slept. When they
Came to share it, Fafner lay in wait
With wicked cunning for his weaker brother,
Because he wished to own the hoard alone.
But Reigen fled, and was seen no more.
The great gods punished Fafner's crime: he
Became the dragon whom you slew today.

In Raupach's *Nibelungen-Hort* the hoard-owning, Nibelung-oppressing giants already existed, and with them at least one way of combining *Das Lied vom Hürnen Seyfrid* and the Scandinavian sources. All that was left for Wagner to do was to add the gods' role in transferring the gold from dwarfs to giants.

At the other end of the scale from the giants are the Nibelungs, from whom the *Mythus* and the *Ring* take their name. Despite the fact that the phrase 'the Nibelungs' was on the lips of the literary world in Wagner's day there was no

---

[50] *Der Nibelungen-Hort*, pp. 18–19.

simple answer as to who the Nibelungs were. The source literature gives such a variety of accounts. In *Das Lied vom Hürnen Seyfrid* Niblung is a dwarf and a king, father of Eugel and his two brothers, the original owner of the hoard. The *Nibelungenlied* sets its own riddle: there too Niblung is a prince, slain by Siegfried, and owner of a hoard guarded by the dwarf Alberich, and the Nibelungs are the prince's retainers. Nibelungs, too, are the inhabitants of Nibelungen-land which Siegfried now rules by right of conquest, and so Siegfried's subjects. But later in the poem the Nibelungs are Gunther and his brothers — the Burgundian kings — and their entourage.[51] The Scandinavian sources know only Gunther and his kinsmen as Hniflungar.

Princes, retainers, hoard-owners, dwarfs; or Siegfried's in-laws and kings: it was a fascinating, if perplexing, subject and one into which Wagner delved deeply. If his final version of the Nibelungs comes very close to Raupach's, it should not mask the other dimensions of his thinking on the subject, visible both in the *Mythus* and in the related essay *Die Wibelungen*.

The subject had naturally roused much interest among scholars, who were keen to solve the mysteries surrounding the Nibelungs' identity. 'The terrible dark forces of the earth and the night, of the subterranean depths and the abyss . . . of the Scandinavian proto- and underworld Niflheim' was von der Hagen's conclusion on the subject of the Nibelungs in his *Die Nibelungen*.[52] Lachmann's description of the genus is similar: 'a superhuman one from the cold, misty realm of the dead'.[53] If this seems a far cry from the pathetic, downtrod-den dwarfs of *Rheingold*, or even Alberich himself, we must remember that in the *Mythus* the underworld, as opposed to merely underground, side of the Nibelungs' nature had been brought out much more strongly: 'Out of the womb of night and death a race emerged which dwells in Nibelheim (Home

---

[51] Aventiure 3, 8, 25 ff.

[52] 'Die furchtbaren finsteren Gewalten der Erde und Nacht, der unterirrdischen Tiefe und des Abgrundes . . . der Nordischen Ur- und Unterwelt Nifl-heim' (pp. 44–5).

[53] 'ein übermenschliches aus dem kalten neblichten todtenreich' ('Kritik', p. 342).

of Mist), that is, in gloomy subterranean chasms and caves: they are called Nibelungs . . .'[54]

Lachmann and von der Hagen also showed that the apparent contradictions in the use of the name Nibelung in the sources could be turned to the advantage of Romantics seeking the essential unity of all being. If the same name was being used for what looked on the surface like different entities, then surfaces were deceptive and at a deeper level the two must be one. Thus von der Hagen, comparing the three dwarf brothers of *Das Lied vom Hürnen Seyfrid* and the Burgundian king Nibelungs of the *Nibelungenlied*, found that 'basically' they were 'one',[55] and while this assessment may not have made much impact on Wagner, since von der Hagen almost invariably found that things were 'basically . . . one', Lachmann also threw his scholarly weight into the argument:

We observe . . . that where the northern saga first establishes the treasure in the power of the dwarfs the South German saga, not without confusion, makes the first lords of the treasure into Nibelungs also, along with Günther and his entourage . . . One can therefore hardly doubt any longer that the former and the latter are of the same race.[56]

The unifying factor between the Nibelungs at the beginning of the story and those at the end, by Lachmann's reckoning, is the hoard; and from an identification of the earlier and later Nibelungs through their common hoard-ownership it was only a small step to suggesting that the name went with the hoard. Such was von der Hagen's argument in another of his contributions to Nibelung scholarship, the introduction to his 1812 Edda edition.[57] Wagner considered this option too; in *Die Wibelungen* he defines the Nibelungs as

---

[54] 'Dem Schoosse der Nacht u. des Todes entkeimte ein Geschlecht, welches in Nibelheim (Nebelheim), d. ist in unterirdischen düstren Klüften u. Höhlen, wohnt: sie heissen Nibelungen . . .' (*Skizzen*, p. 26).

[55] 'im Grunde . . . eins' (*Die Nibelungen*, p. 44).

[56] 'Betrachten wir . . . dass wenn die nordische sage zuerst den schatz in der gewalt der zwerge sein lässt, die süddeutsche nicht ohne verwirrung ausser Günther und seiner umgebung auch die ersten herren des schatzes zu andern Nibelungen macht . . . so wird man schwerlich noch zweifeln, jene und diese sind von einem geschlecht.' ('Kritik', p. 342.)

[57] pp. lxxviii ff.

whoever owns the hoard: 'All endeavour and all striving are
directed towards this hoard of the Nibelungs ... and
whoever possesses it, whoever commands by means of it, is
or becomes a Nibelung.'[58] Further on Wagner explains the
process more fully, incorporating the underworld side of the
Nibelungs' nature and the doom awaiting all those who own
the hoard. The Nibelungs are the 'children of night and
death'; whoever wins their hoard is marked our for death and
joins the Nibelungs in their domain. It is a process from
which Siegfried himself is not exempt:

when Siegfried slew the Nibelung dragon he also won as fair booty
the Nibelung hoard the dragon was guarding. Possession of this
hoard ... is however also the reason for his death, for the dragon's
heir seeks to regain it; he kills Siegfried treacherously ... and
draws him down with him into the gloomy realm of death. *Siegfried himself becomes a Nibelung in consequence.*[59]

Siegfried as Nibelung: it was certainly an ingenious idea;
perhaps too much so for his drama, for in Wagner's *Mythus*
he preferred instead the more modest idea of the Nibelungs
as the first owners and originators of the hoard. In his dramatic sketch he adopted a standpoint closer to that of
Wilhelm Grimm, who had written in his *Heldensage*: 'I do of
course know that it has been suggested that the name passed
on with possession of Nibelung's gold, but I regard this
conjecture as erroneous ...'[60]
    If the Nibelungs were the creators of the hoard it stood to
reason that they were dwarfs.[61] Everyone knew of the
dwarfs' activity in gathering treasure and working precious
metals. Jacob Grimm, for one, writes in his *Mythologie*: 'In

[58] 'Alles Streben und alles Ringen geht nach diesem Horte der Nibelungen ...
und wer ihn besitzt, wer durch ihn gebietet, ist oder wird Nibelung.' (*Schr.*, ii. 119.)
[59] 'als Siegfried den Nibelungendrachen erschlug, gewann er als gute Beute auch
den vom Drachen bewachten Nibelungenhort. Der Besitz dieses Hortes ... ist aber
auch der Grund seines Todes: denn ihn wieder zu gewinnen, strebt der Erbe des
Drachen, — dieser erlegt ihn tückisch ... und zieht ihn zu sich in das finstere Reich
des Todes: *Siegfried wird somit selbst Nibelung.*' (*Schr.*, ii. 132–3.)
[60] 'Ich weiss zwar, dass man die Vermuthung aufgestellt hat, der Name sey mit
dem Besitze von Nibelungs Gold übergegangen, aber ich halte sie für falsch ...' (p.
67).
[61] The word 'Zwerg' ('dwarf') does not appear in the *Mythus*, though Wagner
does use the term 'Schwarzalben' ('black-elves')—a synonym for dwarfs—in *Die
Wibelungen* in connection with the Nibelungs (*Schr.*, ii. 132). See also Ch. 7.

these caves they carry out their activities, accumulate treasure and forge exquisite weapons'.[62] In the Scandinavian hoard saga the treasure originally belonged to the dwarf Andvari, in *Das Lied vom Hürnen Seyfrid* to the dwarf king Nibelung and his sons. Even if the *Nibelungenlied* did not state that the hoard-owning princes Schilbunc and Niblunc were dwarfs, Jacob Grimm did: 'It appears that [the hoard] belonged to dwarfs and that Schilbunc and Niblunc were of elvish [i.e. dwarf] origin.'[63]

Dominant among the Nibelungs is Alberich, and we shall conclude our study of the *Mythus* with the arch-Nibelung himself. Wagner's Alberich, in addition to being a psychologically convincing portrait, is something of a virtuoso performance, for he draws on dwarfs from a host of literary sources and demonstrates the unifying power of contemporary scholarship to perfection.

The common identity of the various dwarf characters which haunt medieval literature was a favourite theme among the scholars. Mone, for instance, in the chapter 'Über den Ages' of his valued *Untersuchungen*, suggests that the dwarfs appearing in the literature under no fewer than six main names and several subsidiaries are one and the same: '[Ages] appears more frequently under his other names. These are: 1) Elbegast, Elegast . . . 2) . . . Alberich, Elberich, i.e., elf lord, elf king . . . Oberon . . . 3) Laurin, Laurein . . . 4) Euglein . . . 5) Malagis . . . Maugis.'[64] This wholesale simplification of the dwarf world cannot just be dismissed as one of the excesses for which Mone was universally criticized (see Chapter 1), for when Wagner discovered the *Altdeutsche Wälder* shortly after completing his *Mythus* he would have found no less a scholar than Jacob Grimm doing much the same thing: in an article entitled 'Über Agges und Elegast' Grimm argues together Agges, Elegast, Elberich, Alberich,

---

[62] 'In diesen höhlen trieben sie ihr wesen, sammeln schätze und schmieden köstliche waffen' (p. 424).

[63] 'Es scheint dass er zwergen angehörte und Schilbunc und Niblunc elbischer art waren.' (*Mythologie*, p. 931.)

[64] 'Er komt häufiger mit seinen andern Namen vor. Diese sind 1) Elbegast, Elegast . . . 2) . . . Alberich, Elberich, d.h. Elfenherr, Elfenkönig . . . Oberon . . . 3) Laurin, Laurein . . . 4) Euglein . . . 5) Malagis . . . Maugis.' (pp. 136–7.)

Oberon, Alpris, Malpriant, Mabrian, and Magis.[65]

The cynic will conclude that for those really determined any dwarf could be found identical with any other. Nevertheless, the scholars conscientiously provided what proof they could for each common identity they postulated, and Wagner would have found inspiration, or at least sanction, for every individual combination manifest in his Alberich.

The starting-point, the dwarf Alberich himself, is the whip-wielding, tarnhelm-wearing guardian of the princes' treasure in the *Nibelungenlied*. That Alberich himself is actually a king, the lord of the dwarfs, is clear from the etymology of the name, as Mone had pointed out (see above), and Lachmann too.[66] Similarly, Jacob Grimm had written of Alberich as king of the dwarfs in the chapter 'Wichte und Elbe' of his *Mythologie*.[67]

If Alberich is king of the dwarfs, then that suggests affinity with Eugel, dwarf king in the *Das Lied vom Hürnen Seyfrid*, hoard-owner and centre of the social conflict with the giants. Eugel's father Nibelung had two further, unnamed, sons, while the *Nibelungenlied* groups together Alberich, Schilbunc, and Niblunc. Wilhelm Grimm discovered here a significant parallel: 'Eugel with his cloak of invisibility and his dwarfs takes the place of Alberich and Nibelung's warriors . . . Schilbung and Nibelung are represented by the two unnamed sons of Nibelung . . .'[68] For von der Hagen, Eugel is simply Alberich under a different name: 'Eugel . . . as Alberich is called in *Das Lied vom Hürnen Seyfrid*'.[69] The bizarrest incident could be used to illustrate the two dwarfs' identity: in a note to the tale 'Dat Erdmännekin' in Grimms' *Märchen* we read: 'The mannekin is Euglin and Alberich, whom the hero similarly only manages to dispose towards himself through force'.[70]

---

[65] J. and W. Grimm, *Altdeutsche Wälder*, i. 31–4.

[66] 'Kritik', p. 342 and n. 9; quoted by Cooke, *World*, p. 125.

[67] pp. 421–2.

[68] 'Euglin mit der Nebelkappe und seinen Zwergen nimmt die Stelle Alberichs und der Nibelungshelden ein . . . Schilbung und Nibelung werden durch die zwei ungenannten Söhne Nibelungs vertreten . . .' (*Heldensage*, pp. 80–1).

[69] 'Eugel . . . wie im Leide von Siegfried, Alberich heisst' (*Die Nibelungen*, p. 79).

[70] 'Das Erdmännchen ist Euglin und Alberich den sich der Held gleichfalls durch Gewalt erst geneigt macht' (iii. 179).

Alberich of the *Nibelungenlied* is intimately involved in a hoard dispute; so too is the Scandinavian Andvari, whose history of river-gold and curse-bearing ring were wanted for Wagner's *Mythus*. Von der Hagen admitted the identity of Alberich and Andvari with the same ease and lack of concern for evidence as before: writing this time in the introduction to his 1812 Edda edition he mentions the occasion 'when Loke fetches the gold from the black dwarfs and Andvari (Alberich, Eugel) . . .'[71]

As for how Alberich came by the gold, the thieving dwarf was so proverbial in the source literature that Jacob Grimm was moved to write in his *Mythologie*: 'All dwarfs and elves are thievish.'[72] Many of the dwarfs in Mone's *Untersuchungen* list and J. Gimm's *Altdeutsche Wälder* paper were renowned thieves, from Ages down. Most likely to have influenced Wagner was Alfrik or Alpris, whom he knew from the *Thidreks saga* as the dwarf who stole the sword Nagelring for Dietrich. The description in the *Thidreks saga* suggests that Alpris was the thieving dwarf *par excellence*: 'This was the dwarf Alpris, the notorious thief and the most cunning of all the dwarfs mentioned in the ancient sagas'.[73] Added to this, the name Alpris or Alfrik has such obvious affinities with the name Alberich that on its second appearance the translator of the *Thidreks saga*, von der Hagen, puts 'Albrich' in the text and adds a footnote suggesting that we read 'Albrich' for 'Alpris' on the previous occasion too.[74] Should Wagner still harbour any doubts that the sword-thief of the *Thidreks saga* was the hoard-guardian of the *Nibelungenlied* they would be removed by Wilhelm Grimm, who declares in the *Heldensage* that the story of Dietrich winning Nagelring from Alpris and the story of Siegfried winning the hoard from Alberich are one and the same.[75]

The last dwarf to concern Wagner was Elberich, royal

[71] 'da Loke das Gold von den schwarzen Zwergen und Andvari (Alberich, Eugel) holt . . .' (p. xlviii; quoted by Cooke, *World*, p. 126).

[72] 'Alle zwerge und elbe sind diebisch.' (p. 434.)

[73] 'Dieses war Zwerg Alpris, der berüchtigte Dieb, und der listigste aller Zwerge, von denen in alten Sagen erzählt wird' (i. 48).

[74] i. 139. Cooke refers to the *Thss* episode as an influence on Alberich's theft of the Rhinegold, and also notes von der Hagen's trans. of the name.

[75] p. 79.

father of the *Heldenbuch* hero Otnit. Here the similarity of
name was felt to be so overwhelming that the identity with
Alberich could not be denied, despite the total lack of correspondence in either role or appearance. According to
Wilhelm Grimm, 'Elberich calls himself a mighty king . . .
He is portrayed as a beautiful child, in contrast to the
Nibelunge Noth [*Nibelungenlied*], where he appears as an old
dwarf with grey beard.'[76] Wilhelm Grimm refers to Elberich
as Alberich elsewhere in his *Heldensage* too,[77] while his
brother Jacob also takes up the argument in the chapter
'Wichte und Elbe' of his *Mythologie.*[78]

The identity of Elberich and Alberich attracted Wagner,
not because he wished to establish any connection with
Elberich's son Otnit but because the prospect of Alberich as
the possible father of a hero interested him. Wagner still
needed a father for Gunther's half-brother Hagen. In the
Thidreks saga Hagen is sired by a dwarf, though the saga does
not say which. Elberich and Alberich, Otnit and Hagen: the
parallel was too neat for Wagner to ignore and too tempting
to resist, and so he duly made Hagen Alberich's son. The
unifying urge of Romantic scholarship was being led to
completion.

Indeed, it is strange that such fulfilment had eluded
Wagner's authorities, though Lachmann, to be sure, appears
to have been on course for a similar discovery in his discussion of Hagen:

In the Scandinavian saga [Hagen] is said to be a Niflung [Nibelung]
and Hniflungr is the name of his son. According to the German
saga his father is a dwarf; or else he is either called Aldrian, in what
sounds to be a foreign adaptation, or, in Eckehard, the cowardly
and loquacious Agazjo, evidently the mythical master thief Agez of
Reimar von Zweter and in the *Titurel.*[79]

---

[76] 'Elberich nennt sich einen mächtigen König . . . Er wird als ein schönes Kind
dargestellt, im Gegensatz zu der Nibelunge Noth, wo er als ein alter, graubärtiger
Zwerg erscheint.' (*Heldensage*, p. 227.)

[77] pp. 236, 342, 391–2.

[78] pp. 421–2.

[79] 'Nach der nordischen sage soll er ein Niflung sein und Hniflungr heisst sein
sohn: nach der deutschen ist sein vater ein alb, oder er heisst theils in einer fremd
lautenden umbildung Aldrian . . . theils bei Eckehard Agazjo, der feige und red-
selige, offenbar der mythische meisterdieb Agez bei Reimar von Zweter und im
Titurel.' ('Kritik', p. 345.)

So Lachmann is arguing that Hagen's father may be Agez; and Agez, Wagner knew from Mone and Grimm, was Alberich.

The circle is complete. With Alberich for his father Hagen too is a Nibelung, and so Wagner, together with von der Hagen and Lachmann, has a Nibelung at either end of his story: as first hoard-owner and as Siegfried's in-law and slayer. Wagner had first introduced the Nibelungs in the *Mythus* as the offspring of night and death; by the end of his drama, with the cold-blooded and relentless Nibelung Hagen, we are beginning to believe him.

6

The Work of 1848, ii: *Siegfrieds Tod*

ALMOST immediately after finishing the *Mythus* Wagner started work on *Siegfrieds Tod*, and by 20 October 1848 the prose plan for all three acts, without the prologue, was complete.[1] The next day Wagner went to the Royal Library and resumed his loans of Nibelung literature, interrupted since 2 October, with the Grimm brothers' *Altdeutsche Wälder*, Ettmüller's *Edda* and the long-sought *Völsunga saga* in von der Hagen's *Nordische Heldenromane* translation. He also borrowed again Jacob Grimm's *Deutsche Rechtsalterthümer*, which he had returned to the library on 2 October. Wagner then resumed work on his opera text. Probably towards the end of October he wrote the prose plan of the Norn Prologue,[2] while the actual poem of *Siegfrieds Tod* in its original version was written between 12 and 18 November 1848.[3]

If the resumption of Wagner's studies leads us to expect widespread changes in the work that followed, then we shall be disappointed. There is only minimal evidence in the poem of *Siegfrieds Tod*, *vis-á-vis* the prose plan, of new influence from the four works Wagner read in the interim. Grimm's *Rechtsalterthümer* was in any case not new reading, and while it was definitely one of the influences on Wagner's text, in *Siegfrieds Tod* Wagner only really expands on what the *Rechtsalterthümer* had already contributed to the prose plan. The scope for the *Altdeutsche Wälder* was limited; apart from the article 'Ueber Agges und Elegast' on dwarf identities, the *Altdeutsche Wälder* had nothing to offer the *Ring*. Ettmüller's *Edda* provided Wagner for the first time with the second half of the 'Heldenlieder', but too late to alter his intention to end

---

[1] *Skizzen*, p. 55.
[2] Ibid.
[3] *Skizzen*, p. 58.

his 'grand heroic opera'[4] at the imposing spectacle of Siegfried's funeral pyre. Even the *Völsunga saga*, by far the most
important of the four sources borrowed for the *Ring*, had to
wait until *Die Walküre* to come into its own.

From the first lines of the Norn Prologue, however, we
can observe one of the most recurrent influences on *Siegfrieds
Tod*, that of Fouqué's *Sigurd*. A meeting of the norns at
Brünnhilde's rock, of which the primary sources know
nothing, had been attempted, prior to Wagner, by Fouqué,
who opens the second Adventure of his dramatic poem with
three norns circling around the sleeping Brünnhilde. As they
dance they fill us in on some of the past events, which in
Fouqué's dramatic poem consist of Brünnhilde's past disobedience, present punishment, and approaching deliverance
in the matter of the dispute between Agnar and Hialmgunnar. At Siegfried's approach they vanish.[5] We suggested in
Chapter 5 that this scene was one of Wagner's main sources
for the history of Brünnhilde's defiance of Wotan, which he
later transferred to the story of Siegmund's death;[6] for
Wagner's norn scene Fouqué provided the model in both
form and function.

Technically, the characteristics of Wagner's norn demi-
goddesses could, with minor exceptions, have come straight
from the Eddas, where there are many references, scattered
and sometimes none too clear-cut, to these female guardians
of fate. In practice, Wagner probably took his information
ready-sorted from writers such as Fouqué and Jacob Grimm
before giving it his own distinctive stamp. The number of
Wagner's norns illustrates this; whereas the number of the
norns in the Eddas is variable, Fouqué, like Wagner, has
three. Fouqué evidently derived his norns from a passage in
the *Snorra Edda* where three norns are named and their function described,[7] for he gives Snorri's names, half-turned to
German as Wurdur, Werdandi, and Skuld, and interprets

---

[4] 'grosse Heldenoper', Wagner's subtitle for the first recension of *Siegfrieds Tod*
(*Skizzen*, p. 58).
[5] Fouqué, *Sigurd*, pp. 61–4; mentioned by Golther, *Grundlagen*, p. 105, as a source
for Wagner's scene.
[6] See ch. 5 nn. 7, 30.
[7] Simrock, *Edda*, p. 252.

them as indicating respective responsibility for past, present and future:

> Wurdur hat das Gerwordne gelenkt,
> Werdandi lenkt das Werdende jetzt,
> Und Skuld had Kunde, was kommen soll.[8]

> Wurdur directed what came about,
> Werdandi steers what is happening now,
> And Skuld has knowledge of what is to come.

Either from Fouqué, then, or from Grimm's discussion of the same matter in the chapter 'Weise Frauen' of his *Mythologie*,[9] Wagner found the inspiration for the time-orientated distribution of his norns' dialogue in *Siegfrieds Tod* and *Götterdämmerung*.

Wagner does not name the norns. That he had the accredited norn names, as used by Snorri and Fouqué, somewhere in mind, however, is clear from a line the first norn speaks towards the close of the scene:

> Meinem Brunnen nahet sich Wotan.[10]
> Wotan approached my well.

She is evidently Fouqué's Wurdur or the Urdr of the *Snorra Edda*, whose name, as Grimm points out, appears again in 'Urd's well' under the root of the world ash Yggdrasil: 'Even in the north Urdr must have been more important than the other two [norns], for the well by the sacred ash is called 'Urd's well' after her . . .'[11]

As compensation for omitting the norns' names Wagner included information which was probably not apparent to Fouqué concerning their ages. In the stage direction at the beginning of the *Siegfrieds Tod* prologue Wagner announces the norns as follows: 'The first (eldest) attaches the line firmly to a fir-tree on the extreme right. The second

---

[8] *Sigurd*, p. 62.
[9] pp. 376–7.
[10] *Schr.*, ii. 170.
[11] 'Selbst im Norden muss *Urdr* bedeutsamer als die beiden andern gewesen sein, denn der brunnen an der heiligen esche heisst nach ihr *Urdarbrunnr* . . .' (*Mythologie*, p. 379).

(younger) winds it round a stone, left. The third (youngest) holds the end, downstage centre.'[12] Here Wagner was drawing on Jacob Grimm, whose *Mythologie* appeared long after Fouqué's dramatic poem. Taking a passage from *Norna-gest* as his starting-point, Grimm postulates the norns' different ages: 'The third one, that is, Skuld, is called "the youngest"; they were therefore presumed to be of different ages, with Urdr as the eldest.'[13]

Grimm was the source, too, of one of Wagner's great improvements over Fouqué: the stage business. Whereas Fouqué's norns are engaged in an aimless-seeming dance, Wagner's are busy with the web of fate. Grimm had assembled numerous references to spinning and weaving norns and other related ladies, including the important passage in *Helgakvidha Hundingsbana fyrri*, on which he comments:

In this remarkable passage it is related that norns entering the castle by night wound the threads of fate for the hero and spread out the golden line right across the sky; one norn secured an end of the thread in the east, the second in the west, the third fastened it in the north.[14]

Out of this Wagner devised the business with the line which accompanies the whole scene through to the closing lines—

Schliesset das Seil, wahret es wohl!
Was wir spannen, bindet die Welt.[15]

Draw in the line, preserve it well!
What we have spun binds the whole world.

---

[12] 'Die Erste (Älteste) knüpft das Seil, zur äussersten rechts, an einer Tanne fest. Die Zweite (Jüngere) windet es links um einen Stein. Die Dritte (Jüngste) hält das Ende in der Mitte des Hintergrundes.' (*Schr.*, ii. 167.)

[13] 'Diese dritte, folglich *Skuld*, heisst "die jüngste", sie wurden also von verschiednem alter, und *Urdr* als die älteste angenommen.' (*Mythologie*, p. 381.)

[14] 'In dieser merkwürdigen stelle ist gesagt, dass nachts in die burg tretende nornen dem helden die schicksalsfäden drehten und das goldne seil mitten am himmel ausbreiteten; eine norn barg ein ende des fadens gen osten, die andere gen westen, die dritte festigte gegen norden.' (*Mythologie*, pp. 379–80.) Grimm actually quotes the passage from the poem, but only in the original Old Norse. However, Wagner knew the *Helgakvidha* in the Grimm brothers' German trans. in their *Edda*. Golther (*Grundlagen*, p. 104) suggests it as a source without ref. to the *Mythologie*. Further 'spinning and weaving' refs. are found in the same ch. of the *Mythologie*, pp. 385–7.

[15] *Schr.*, ii. 179.

He also took over the three points of the compass from the *Helgakvidha*:

> *Die erste Norn*
> In Osten wob ich.
> *Die Zweite*
> In Westen wand ich.
> *Die Dritte*
> Nach' Norden werf' ich.[16]

> *First norn*
> I wove in the east.
> *Second norn*
> I wound in the west.
> *Third norn*
> I cast to the north.

Despite the radical rewriting of this scene in *Götterdämmerung*, whereby not one line remained of the original text, Wagner retained the 'spinning and weaving' motif and even integrated it more closely into the action of his drama. He also managed to rescue one of the cardinal points as the third norn returns the rope:

> von Norden wieder
> werf ich's dir nach:
> spinne, Schwester, und singe![17]

> From the north I cast it
> back to you;
> spin, sister, and sing!

With the disappearance of the norns from the stage Siegfried and Brünnhilde enter. Siegfried, who only yesterday vowed eternal love to Brünnhilde, is leaving. None of the primary sources contain a farewell scene, nor indeed do they explain why Siegfried, having just given his word to Brünnhilde, should promptly abandon her. It was again Fouqué who first suggested that Siegfried was forced to leave by his duty as a hero. Fouqué does not give a farewell scene as such, but the germ of parting is present already at the waking of Brünnhilde:

---

[16] *Schr.*, ii. 167.
[17] *Schr.*, vi. 180.

Brynhildur
Von meinem Lager fort
Wird in die Welt hinaus dein kühner Sinn
Dich treiben, neuen Abenteuern nach . . .[18]

Brynhildur
Forth from my bed
Into the world beyond your brave heart
Will drive you, in search of new adventures . . .

This was undoubtedly the main inspiration for the *Siegfrieds Tod* parting lines:

Brünnhilde
Zu neuen Taten, teurer Helde,
wie liebt ich dich—liess' ich dich nicht?[19]

Brünnhilde
To fresh deeds, dearest hero—
how could I love you, and not release you?

Perhaps, too, Wagner was drawing on Göttling, whose *Nibelungen und Gibelinen* we think can be shown to have influenced the composer (see Chapter 4). Göttling's tract includes a somewhat free résumé of the Nibelung story from the Edda: which Edda, Göttling does not specify, and some of what he includes does not appear to derive from any known Edda. In this instance he writes of the lovers that Brünnhilde 'finally became Sigurd's wife. But Sigurd's desire for action began to stir again; he was drawn forth into the wide world.'[20] Göttling was adding nothing new to what Wagner had learnt from Fouqué, but he perhaps helped reinforce it.

When Göttling writes of Siegfried and Brünnhilde that she 'finally became his wife' he is again venturing on ground circumvented by the primary sources. Those sources which follow the Scandinavian tradition of a meeting between Siegfried and the valkyrie before he wins her for Gunther end the episode with nothing more substantial than exchanged

---

[18] *Sigurd*, p. 70.
[19] *Schr.*, ii. 170.
[20] 'ward letzlich sein Gemahl. Aber Sigurds Trieb nach Thaten ward wieder rege; es zog ihn fort in die Weite.' (*Nibelungen und Gibelinen*, p. 48.)

promises between them.[21] Against this, Wagner was quite positive that the two were man and wife, expressing it unequivocally in Brünnhilde's opening—

> doch meiner Stärke magdlichen Stamm
> nahm mir der Held, dem ich nun mich neige[22]

> but the hero, to whom I now bow,
> took my maidenhead, source of my strength

—and continuing through to her proud 'I am his wife' of Act II.[23] Once more it is likely that Göttling's influence worked hand in hand with that of Fouqué, who also makes it quite clear that relations between Siegfried and Brünnhilde are progressing beyond promises with references to 'bride', 'bridegroom', 'wedding cup', and 'bridal chamber'.[24]

These two items of supposed influence from Göttling's *Nibelungen und Gibelinen* do not in themselves constitute a case for Wagner having known the work, since both merely reduplicate what he already knew from Fouqué. However, we should not simply dismiss them as redundant, for later in *Siegfrieds Tod* we shall find further evidence, harder to refute, that Göttling was an active influence on the *Ring*.

The reception party of Gibichungs who await Siegfried's arrival at the opening of Act I and the challenge Siegfried issues once he is there, decorously condensed by Wagner to

> nun ficht mit mir—oder sei mein Freund![25]

> now fight with me—or be my friend!

—could both equally well be taken direct from the *Nibelungenlied* as from Fouqué, who opens Adventure 3 of his dramatic poem in the same manner. The horse, however—or rather Siegfried's manifest concern for it—can only have been

---

[21] Aslaug, the daughter of Siegfried and Brünnhilde who appears late in the *Vs* and *SnE*, owes her existence to considerations outside the saga proper. It was not lost on Wagner's contemporaries that she is primarily a device to link the house of Ragnar Lodbrok, whose saga follows the *Vs* immediately in the MS, with the famous dragon-slayer; see von der Hagen, *Vs*, pp. xii–xv.

[22] *Schr.*, ii. 170.

[23] 'Sein Gemahl bin ich' (*Schr.*, ii. 225).

[24] 'Braut', 'Bräutigam', 'Hochzeittrank', 'Brautgemach' (*Sigurd*, pp. 68–70).

[25] *Schr.*, ii. 178.

inspired by Fouqué. None of the primary sources record any words of Siegfried on the subject of the care of his horse, nor any particular solicitude evinced. It was certainly part of a hero's duty after a long day in the saddle to look first to the welfare of his steed; Wagner's Siegfried, however, arrives by boat, not on horseback, which makes the priority he gives to Grane on arrival all the more remarkable. He has barely drawn breath since landing before he is asking:

> Wo berg' ich das Ross?
>
> Where shall I keep my horse?

and when Hagen offers to stable him he is admonished:

> Wohl hüte mir Grane! Du hieltest nie
> von edlerer Zucht am Zaume ein Ross.[26]
>
> Guard Grane well! You never held
> the rein of horse more nobly bred.

Wagner obviously had in mind the scene of Sigurd's arrival in Fouqué's dramatic poem. Since Sigurd is here reported as riding towards the company, it comes as no surprise that his first words after greeting his hosts are:

> Ist wer dabei,
> Der mir mein treues Ross zur Wartung abnimmt?
> (es treten einige Diener vor)
> Ihr lieben Leute, nehmt dies Pferd in Acht,
> Behandelt's höflich, sonsten wird es bös',
> Denn edler Gattung ist's, heischt feine Zucht.[27]
>
> Is someone there
> Who'll take my faithful horse and care for him?
> (some servants come forward)
> Good people, pay attention to this horse,
> Treat him civilly, else he'll grow bad-tempered,
> For he's of noble stock, demands good breeding.

Whenever it was that Wagner read Fouqué's *Sigurd* this scene evidently made an indelible impression on him.

---

[26] Ibid.
[27] *Sigurd*, p. 95; Golther, *Grundlagen*, p. 106, also notes the parallel.

We suggested in Chapter 5 that Wagner was initially introduced to the magic potion by Fouqué who, sticking faithfully
to the *Völsunga saga* original, lets the potion be master-
minded and administered by Grimhildur, the Gibichungs'
mother. If Wagner chose to depart from Fouqué in this
respect and place the responsibility with Hagen and Gutrune,
he was nevertheless influenced by Fouqué's extremely effective handling of the potion as he demonstrates the drug
working on Sigurd's memory. Immediately after first tasting
the liquid Sigurd is seized with a sense of loss and cries out:

> *Sigurd*
>   Ich habe was verloren.
> *Grimhildur*
> Von deinem Schatz doch nicht?
> *Sigurd*
>   Aus den Gedanken. —
> Noch eben erst konnt' ich mich d'rauf besinnen,
> Und's war mir lieb, im tiefsten Herzen lieb.
> Mit einemmal entfiel's den Sinnen, fiel
> Als wie in's bodenlose Meer hinein.
> Ich irr' am Ufer—lasst mich suchen, bitt' Euch.[28]

> *Sigurd*
>   I've lost something.
> *Grimhildur*
> Not from your treasure, surely?
> *Sigurd*
>   From my mind. —
> A moment past I could remember it,
> And I loved it deep within my heart.
> Then all at once it slipped my consciousness,
> Fell as though into the groundless sea.
> I stray on shore—let me seek, I beg you.

We witness his efforts to recall the places and people he has
just left: his encounter with Brünnhilde—until, persuaded by
Grimhildur, he empties the cup, whereupon the struggle
ceases.[29] The pathos of Fouqué's scene left no trace in *Siegfrieds Tod* but, to anticipate a little, in the rewriting for *Götterdämmerung* it bore fruit in the arousal of some dim but

---

[28] *Sigurd*, pp. 105–6.
[29] *Sigurd*, pp. 106–7.

insufficient recollection in Siegfried at the description of the way to Gunther's valkyrie bride.[30]

One further point on the potion: as Siegfried raises the cup to his lips he dedicates the draught to Brünnhilde with the words:

> den ersten Trunk zu treuer Minne,
> Brünnhilde, trink' ich dir![31]
> The first draught in faithful remembrance,
> Brünnhilde, I drink to you!

Wagner's use of the word 'Minne' has evidently caused some confusion, for it appears in various English editions of *Götterdämmerung* translated as 'love'.[32] What Wagner had in mind, however, was more likely the ceremony of 'Minnetrinken' as described by Jacob Grimm in the *Mythologie*: 'It was the custom to honour someone who was absent or deceased by mentioning him at gatherings and banquets and draining a cup in his memory. This draught was called "erfi dryckja" in Old Norse, and also "minni"'.[33] 'Minnetrinken' is thus a drink in memory of absent friends, and 'Minne' in this context is used, not in its archaic sense of 'love', but in its even more archaic sense of 'remembrance'. The irony Wagner intended in these lines is therefore not that Siegfried ceases to love Brünnhilde but that in drinking to her memory he should forget her.

Hard upon the 'Minnetrinken' follows the blood-brotherhood ceremony, which in *Siegfrieds Tod* is described like this: 'Hagen fills a drinking-horn with fresh wine. Siegfried and Gunther scratch their arms with their swords and hold them for a short while over the drinking-horn.' After the oath:

---

[30] *Schr.*, vi. 195–6.

[31] *Schr.*, ii. 180.

[32] e.g. in Andrew Porter's trans. (Folkestone, 1976), or William Mann's for the Friends of Covent Garden series (London, 1964). Wagner's *Götterdämmerung* lines, slightly amended from *Siegfrieds Tod*, run: 'den ersten Trunk | zu treuer Minne, | Brünnhilde, | bring' ich dir!' This becomes respectively 'this drink, the first | I taste as lover, | Brünnhild, I drink to you!' (Porter, *The Ring*, p. 292) and 'this first drink | to true love, | Brünnhilde, I offer it to you.' (Mann, *Götterdämmerung*, p. 27.)

[33] 'Einem abwesenden oder verstorbenen pflegte man zu ehren indem man seiner bei versammlung und mahlzeit erwähnte, und auf sein andenken einen becher leerte, dieser trunk wurde altn. *erfi dryckja*, und wiederum *minni* genannt.' (pp. 52–3.)

'They drink in turn, half each; then Hagen . . . dashes the horn to pieces. Siegfried and Gunther clasp hands.'[34]

Strangely enough for a procedure which sounds so thoroughly Germanic, Wagner's blood-brotherhood ceremony was not sanctioned by any of his authorities. The Eddas and *Völsunga Saga* mention oaths sworn between the in-laws but fail to give any details; Fouqué's heroes are content with just the handshake.[35] Closest to Wagner is Simrock's description in the *Amelungenlied* of the blood-friendship ceremony between Etzel and Alpker:

> Den Blutsbecher leert' er mit ihm beim jüngsten Schmaus,
> Und trank mit dem eignen vermischt des Freundes Blut.[36]

> He drank the blood-cup with him while at the latest feast,
> And drank his friend's blood mixed in with his own.

However, since this occurs in the third volume of Simrock's epic, which was not published until 1849, Wagner evidently did not draw on it for the *Siegfrieds Tod* of 1848. Similarly, we should probably discount Loge's appeal to Odin in the *Oegisdrecka*:

> Gedenkt dir, Odin,
> Wie wir in Urzeiten
> Das Blut mischten beide?[37]

> Remember, Odin,
> How in ancient times
> We both mixed blood?

—since in all probability Wagner did not read the poem until it appeared in Simrock's translation in 1851.

The only scholar who could offer Wagner any help was Jacob Grimm. Overall Grimm had quite a lot to say on blood-brotherhood ceremonies. In his *Deutsche Sprache* he includes various accounts, mostly in Greek. Coming nearer to home, in the chapter 'Symbole' of his *Deutsche Rechtsalter-*

---

[34] 'Hagen füllt ein Trinkhorn mit frischem Wein. Siegfried und Gunther ritzen sich mit ihren Schwertern die Arme und halten diese eine kurze Weile über das Trinkhorn. . . . Sie trinken nacheinander, jeder zur Hälfte; dann zerschlägt Hagen . . . das Horn; Siegfried und Gunther reichen sich die Hände.' (*Schr.*, ii. 182.)

[35] *Sigurd*, p. 107.

[36] *Heldenbuch*, vi. 160.

[37] Simrock, *Edda*, p. 53.

thümer he writes under the subheading 'Blut': 'Solemn oaths, vows and alliances were confirmed with blood . . . On entering into brotherhood both parties let their blood run into a hole together so that it became mixed with earth . . .'[38] No mixing of blood with wine, then, and no drinking the result. Eventually, however, Grimm does produce a description of a blood-brotherhood ceremony similar in outline to Wagner's, but only to question its authenticity: 'No German saga mentions symbolic blood-drinking, the mixing of blood with wine, unless one attributes what the *Gesta Romanorum* narrates about a friendship bond in chapter 67 to German usage.'[39] Curiously, if Wagner took the trouble to check Grimm's reference to the *Gesta Romanorum* in the translation by Grässe which he had at home he would have found little correspondence between the two accounts: there each of the two friends drinks blood from the other's right arm, but there is no mixing of the blood and no wine.[40] To conclude, if Grimm felt it necessary to dispute this version of the ceremony in his book and Simrock used it in volume iii of his *Amelungenlied*, the idea was obviously current at the time and Wagner was evidently acquainted with it.

As guarantor of their brotherhood vows Siegfried and Gunther invoke Wotan:

> Wotan, weihe den Trank,
> Treue zu trinken dem Freund!
> Waltender, wahre den Eid
> heilig einiger Brüder![41]

> Wotan, bless the wine
> to drink faith to the friend!
> Commander, watch over the oath
> of brothers' sacred union!

---

[38] 'Feierliche eide, gelübde und bündnisse wurden mit blut bekräftigt . . . bei eingehung der brüderschaft liessen beide freunde ihr blut in eine grube zusammenrinnen, dass es sich mit der erde vermischte . . .' (p. 192).

[39] 'Des symbolischen bluttrinkens, der mischung des blutes mit wein thut keine deutsche sage meldung, oder man müste, was die gesta roman. cap. 67 von einem freundesbunde erzählen, auf deutsche gewohnheit zurückführen.' (*Rechtsalterhümer*, p. 193). Golther suggests that this was Wagner's source anyway. (*Grundlagen*, p. 106).

[40] p. 120.

[41] *Schr.*, ii. 182.

Again we suspect Fouqué's influence:

> *Gunnar*
>> so mir Odin hilfreich sei,
> Freudvoll mein Leben, schmachesfrei mein Tod,
> Gelob' ich, Sigurd, dir Genossenschaft.[42]

> *Gunnar*
>> so help me Odin,
> Joyful in life and free of shame in death,
> Sigurd, I promise fellowship with you.

Alternatively or additionally, Wagner may have been inspired by another authority from the 'possibles' list, Wilhelm Müller. In the all-important chapter 'Odhinn (Wuotan, Wodan)' of his *Altdeutsche Religion* Müller writes of Wotan as protector of oaths generally and of his association with brotherhood vows in particular: 'The god watches specially . . . over the sanctity of oaths. This is why the Scandinavian custom of drinking blood brotherhood was attributed to him.'[43]

Perhaps Wagner had Müller also in mind at his other oath scene, Siegfried's oath of purgation in Act II, which too begins with an appeal to Wotan:

> Wotan! Wotan! Wotan!
> Hilf meinem heiligen Eid![44]

> Wotan! Wotan! Wotan!
> Aid my sacred oath!

But Wagner's chief mentor on oaths in legal processes was Jacob Grimm, whose *Deutsche Rechtsalterthümer*, in which Wagner was currently immersed, contains a chapter on oaths. One thing Grimm did for Wagner was to broaden the field, for he refers to the practice of calling on gods generally, not only Wotan in particular, at the swearing of oaths, and notes that Donar and Wuotan were the preferred gods in Germany.[45] The *Siegfrieds Tod* purgation oath contains a

---

[42] *Sigurd*, p. 116.
[43] 'Insbesondere wachte der gott . . . über die heilighaltung des eides. Daher wurde die nordische sitte blutsbruderschaft zu trinken auf ihn zurückgeführt.' (pp. 192–3.)
[44] *Schr.*, ii. 203.
[45] *Rechtsalterthümer*, pp. 894–5.

veritable pantheon: Siegfried invokes Wotan, and Brünnhilde retaliates by calling on the 'glorious goddess':

> Höre mich, herrliche Göttin!
> Hüterin heiliger Eide![46]
>
> Give ear to me, glorious goddess!
> Guardian of sacred oaths!

—presumably Fricka—while the crowd invites Donner to intervene:

> Hilf Donner! Tose dein Wetter
> zu schweigen die wütende Schmach![47]
>
> Help, Donner! Let your storm roar
> to silence the raging disgrace!

Grimm further helped with the oath procedure. In Wagner's drama Siegfried swears the oath with his right hand on Hagen's spear.[48] Grimm does not mention a spear, but he does say that oaths might be sworn on a sword and gives two interesting reasons why: 'Either because the sword was sacred to a god (Freyr? Tyr?), or in order to indicate that it should strike down a perjuror.'[49] As so often, Wagner's genius manifests itself in contriving to have it both ways: by substituting the spear for Grimm's sword he has Siegfried swear on the weapon which is sacred to the god he is invoking and which in Act III will avenge his perjury—or at least, that is Hagen's argument:

> *([Hagen] stösst seinen Speer in Siegfrieds Rücken . . .)*
> *Gunther und die Mannen*
> Hagen, was tust du?
>
> .   .   .   .
>
> *Hagen*
> *(auf den zu Boden Gestreckten deutend)*
> Meineid rächt' ich an ihm![50]

---

[46] *Schr.*, ii. 203.
[47] Ibid.
[48] Ibid.
[49] 'Entw. weil das schwert einem gott (Freyr? Tyr?) geheiligt war oder damit anzuzeigen, es solle den meineidigen treffen.' (*Rechtsalterthümer*, p. 896.)
[50] *Schr.*, ii. 221.

([Hagen] plunges his spear into Siegfried's back . . .)
Gunther and the warriors
Hagen! What's happening?

.   .   .   .

Hagen
(pointing to Siegfried lying stretched out on the grass)
I've avenged perjury on him!

Siegfried's oath is followed by the dramatic moment when
Brünnhilde wrests his hand from the spear and makes her
counter-oath. Grimm, again in the *Deutsche Rechtsalterthümer*,
writes that the opposing party, if it did not believe the oath,
could interrupt it, generally by taking the swearer's hand
away from the sacred object on which the oath is being
sworn and also by offering to swear a more powerful oath.[51]

More gods feature in the sacrifices Hagen organizes to
celebrate the home-coming of Gunther and his bride in Act II:

Starke Stiere sollt ihr schlachten,
am Weihstein fliesse Wotan ihr Blut!

.   .   .   .   .   .

Einen Eber fällen sollt ihr für Froh,
einen stämmigen Bock stechen für Donner;
Schafe aber schlachtet für Fricka,
    dass gute Ehe sie gebe![52]

Strong-built bulls I bid you slaughter,
their blood shall flow on the altar to Wotan!

.   .   .   .   .   .

A boar should then be brought down for Freyr,
and stick a sturdy goat for Donner;
slaughter sheep, however, for Fricka,
    to grant a good marriage!

For his sacrifices Wagner reverted to his old vade-mecum,
that other work of Jacob Grimm's, the *Mythologie*. In his
chapter 'Gottesdienst' ('Worship'), Grimm lists the principal
sacrificial animals: after horses come cattle, boars, rams, and
goats.[53] Further on, in the chapter 'Bäume und Thiere'

---

[51] *Rechtsalterthümer*, pp. 904–5.
[52] *Schr.*, ii. 197.
[53] pp. 44–6.

('Trees and Animals'), he picks out the boar as the sacrificial animal dedicated to Freyr and the goat as sacred to Thor.[54] Two of Wagner's dedications were thus accounted for. It is less clear why he should have connected bulls with Wotan or sheep with Fricka. The association does not appear to be known in the sources; on Wotan, for instance, Wilhelm Müller writes: 'We do not know what particular animal sacrifices were made to the god'.[55] Perhaps Wagner was simply allocating animals remaining from Grimm's 'Gottesdienst' list, though the idea of sheep for Fricka does suggest some kinship with the ram-chariot he concocted for the goddess in *Die Walküre*, again without direct precedent in the sources.[56]

The curtain rises in Act III of *Siegfrieds Tod* with the Rhinemaidens petitioning 'Mistress Sun' for the return of their gold.[57] Again, Jacob Grimm is the source of Wagner's poetic image, for he writes in his *Mythologie*: 'Until quite recent times people speaking of the sun or moon favoured the expressions "Mistress Sun", "Sir Moon"'.[58]

The chapter on elements in the same source provided Siegfried's symbolic action later in the scene, which underlines his refusal to be frightened into parting with the ring:

> *Siegfried*
> Fasste er nicht meines Fingers Wert,
>   den Reif geb' ich nicht fort:
>   denn das Leben—seht:—so—
>   werf' ich es weit von mir!
> *(Er hat mit den letzten Worten eine Erdscholle
> vom Boden aufgehoben und über sein Haupt hin-
> ter sich geworfen.)*[59]

---

[54] pp. 631–2.

[55] 'Welche thieropfer dem gotte besonders gebracht wurden, wissen wir nicht' (*Altdeutsche Religion*, p. 209).

[56] Golther suggests that Fricka's ram-chariot derives from a mention in Grimm's *Mythologie* (p. 304) of a ram-wagon for Thor instead of the more usual goat-drawn one (*Grundlagen*, p. 59).

[57] 'Frau Sonne': *Schr.*, ii. 210.

[58] 'Das volk pflegte sich bis auf die spätere zeit, von sonne und mond redend, gern auszudrücken "frau sonne", "herr mond"' (p. 666).

[59] *Schr.*, ii. 215.

Siegfried
Though it weren't worth the finger it's on,
I'll not give up the ring:
for life — look: — thus —
do I toss it away from me.
(At the final words he has lifted a clod of earth from the ground and thrown it behind him over his head.)

According to Grimm, 'Our sixteenth-century mercenaries still threw a clod of earth on going into battle . . . as a sign of their complete detachment from life.'[60]
'Here we learn', according to Wagner, 'that Siegfried is infinitely wise, for he knows the highest thing of all: that death is better than a life of fear . . .'[61] Siegfried clings to the fateful gold. Yet no one could accuse him, unlike other would-be ring-owners, of gold-lust; he has, after all, retained only the ring and the tarnhelm from the immense hoard which was at his disposal. It is a question of principle, Wagner's reduced-scale demonstration of the ethos which in his sources causes Siegfried to take the gold despite the warning attached. In the Edda the attitude is predominantly fatalistic and laconic:

Sigurd
Goldes walten
Will ein Jeder
Stäts bis an den Einen Tag.
Den Einmal muss
Doch jeder Mann
Fahren von hinnen zu Hel.[62]

Sigurd
All men want
To wield gold
Right until their final day.
For every man
Must surely sometime
Travel from here to hell.

[60] 'Noch unsere landsknechte des 16 jh. warfen, in die schlacht gehend, eine erdscholle . . . zum zeichen aller lossagung von dem leben.' (*Mythologie*, p. 609; quoted by Golther, *Grundlagen*, p. 108.)
[61] 'Hier erfahren wir, dass Siegfried unendlich wissend ist, denn er weiss das Höchste, dass Tod besser ist, als Leben in Furcht . . .' (letter to August Röckel, 25 Jan. 1854, *Dokumente*, p. 93).
[62] *Fafnismál*, ver. 10: Simrock, *Edda*, p. 162.

Fouqué and Raupach had already reworked the Eddic fatalism into a more positively heroic stand against the power of fear. Fouqué's Sigurd reflects:

> Doch einmal ist der reiche Hort nun mein,
> Und gar ein kläglich Stücklein dünkt es mich,
> Um Drohung seinem Eigenthum entsagen.[63]

> But now indeed the wealthy hoard is mine,
> And it seems a feeble show to me
> To give up what one owns because of threats.

In Raupach's adaptation of the falcon dream motif from the *Nibelungenlied* we read:

> *Chriemhild*
> Mich dünkt, der Tod liegt schlafend auf dem Hort;
> Willst Du ihn wecken?
>   *Siegfrid*
>   Ja, ich will's, und wär' es
> Der Teufel selbst, ich weckt' ihn.
>   *Chriemhild*
>   O gedenke
> Des Edelfalken!
>   *Siegfrid*
>   Nun, ich thu's, und denke,
> Ein edler Falke darf die Furcht nicht kennen.[64]

> *Chriemhild*
> I fancy Death lies sleeping on the hoard.
> Would you wake him?
>   *Siegfrid*
>   I want to, yes, and were it
> The devil himself I'd wake him.
>   *Chriemhild*
>   O remember
> The noble falcon!
>   *Siegfrid*
>   Well, I do, and fancy
> A noble falcon should not know what fear is.

Something of the same heroic ethos spills over into the

---

[63] *Sigurd*, p. 59.
[64] *Der Nibelungen-Hort*, pp. 20–1.

handling of the invulnerability motif. Siegfried's horny skin, which he acquires in the German sources by bathing in the dragon's blood or molten scales, was a mixed blessing to the nineteenth-century dramatist. On the one hand it tends to downgrade Siegfried's heroism and invite further intrusion from the burlesque element which is already quite sufficiently represented in his tales. On the other hand, the flaw in Siegfried's horn protection, where a linden leaf fell between his shoulders or, in the *Thidreks saga*, where he simply could not reach, opens the way for the 'Achilles' heel' topos and the theme of betrayal which is so potent in the story of Siegfried's death.

Wagner decided he could claim the benefits and escape the penalties of the invulnerability motif if he abandoned the horny skin and let Siegfried's protection be conferred by the runic knowledge Brünnhilde possesses in the Scandinavian sources. However, the substitution of charms for dragon blood ruled out the intervention of the linden leaf and Wagner had to cast around for an alternative reason why Siegfried's back was left unprotected. The solution he adopted has the merit of paying full tribute to Siegfried's bravery:

> *Brünnhilde*
> niemals, das wusst' ich, wich' er dem Feind,
>   nie reicht' er ihm fliehend den Rücken,
>   an ihm drum spart' ich den Segen.[65]

> *Brünnhilde*
> I knew he would never give ground to the foe,
>   never turn him his back in flight,
>   so I spared it the spell's protection.

It is also a direct echo in both sentiment and idea of Raupach, who likewise felt the need to 'heroize' the horny skin and whose Siegfrid in consequence, when warned of the flaw in the protection of his back, responds coolly:

> Mag seyn: da trifft mich wenigstens kein Feind.[66]
> May be: at least no foe will reach me there.

And so Siegfried goes on to his death, forewarned, even

---

[65] *Schr.*, ii. 205.
[66] *Der Nibelungen-Hort*, p. 21.

forearmed, but deliberately heedless. We mentioned when discussing the *Mythus* that Wagner's death scene is based primarily on the *Nibelungenlied*. Externally, the feature which marks the *Siegfrieds Tod* scene out from its *Nibelungenlied* model is the return of Siegfried's memory at the point of death and with it his ecstatic recollection of Brünnhilde. Wagner achieves the effect by means of a memory-restoring potion Hagen gives Siegfried to drink, the antidote evidently to the forgetfulness-brew of Act 1:

> *Hagen*
> Trink erst, Held, aus meinem Horn!
> Ich würzte dir holden Trank,
> die Erinnerung hell dir zu wecken,
> das Fernes nicht dir entfalle.[67]

> *Hagen*
> Drink first, hero, from my horn!
> I've seasoned a special brew
> to rouse your memory clearly,
> so that distant things don't escape you.

The memory-restoring draught does not come from the Siegfried literature. Perhaps Wagner was thinking of an episode in the *Sörlathattr*, quoted in Frauer's book on valkyries. In the *Sörlathattr* the hero Hethin, having sworn brotherhood with Högni, kills Högni's wife and carries off his daughter after his memory has been tampered with by a drink which the valkyrie Göndul offers him. After the atrocities Hethin falls asleep; when he awakes the valkyrie is gone but his memory is restored.[68]

Whether or not the *Sörlathattr* episode had any bearing on Wagner's drama, the memory-restoring potion necessitated Siegfried's drinking from a horn; natural enough normally, this runs quite counter to the *Nibelungenlied* version of the saga, where an elaborate conspiracy of salted food, delayed wine, and race through the forest is evolved solely to induce Siegfried to drink from a stream, where his back will present an easy target for Hagen's spear. However, the illustrators of Pfizer's edition of the *Nibelungenlied*, Schnorr von Carolsfeld

---

[67] *Schr.*, ii. 220.
[68] Frauer, *Die Walkyrien*, pp. 23–30.

and Neureuther—the same who gave us the Rhinemaidens—
removed all obstacles posed by the text, for with blithe indif-
ference to the exigencies of the poem they represent Siegfried
at the critical moment drinking from a horn while Hagen
lunges at his back.[69]

As Siegfried lies dying memory of Brünnhilde floods back:

> Brünnhild! Brünnhild!
> Du strahlendes Wotanskind!
> Hell leuchtend durch die Nacht
> seh' ich dem Helden dich nah'n.[70]

> Brünnhilde! Brünnhilde!
> O radiant Wotan-child!
> Brightly shining through the night
> I see you draw near to the hero.

It is an excellent way to die, the mood jubilant, the shadows
of misunderstanding finally dispelled. One cannot help but
feel that the primary sources, where at best the wife Siegfried
dies thinking of is Gutrune/Kriemhild, rather missed out
here. Certainly Wagner's standard authorities on memory-
loss:—the *Völsunga saga* and Fouqué—failed him, for they
restore Siegfried's past to him as soon as his mission to win
Brünnhilde for Gunther is accomplished and thus rule out
any dramatic death-bed recollection.[71]

Inspiration may have come to Wagner from another
quarter, however. We mentioned earlier Göttling's retelling
of the Eddic Siegfried story in his *Nibelungen und Gibelinen*; at
the point where Siegfried loses his memory Göttling's version runs into difficulties, for the Eddas are not forthcoming
on the subject of Siegfried's loss of memory and are quite
silent on its return. The memory-robbing potion Göttling
takes from the *Völsunga saga*,[72] but rather than stay with the
Völsunga saga for the restoration of Siegfried's memory Göttling goes his own way. First he makes a deliberate break
with the older tradition by stating that on his return to the
Gibichungs' hall Siegfried does *not* remember Brünnhilde:

[69] *Der Nibelungen Noth*, p. 182 (see jacket illustration).
[70] *Schr.*, ii. 221.
[71] *Vs*, p. 132; Fouqué, *Sigurd*, pp. 153–4.
[72] *Nibelungen und Gibelinen*, p. 48.

'Brunhild . . . followed Gunther quietly and became his wife in Burgundy, and Siegfried still did not recognize her.'[73] Instead, Göttling has the happy inspiration that Siegfried's memory should return after he has received his death-wound, when 'as the blood rushed from his heart the remembrance of his lost happiness with Brunhild returned.'[74]

Like Wagner, Göttling restores Siegfried's memory at the moment of death; like Wagner, Göttling has Siegfried die with Brünnhilde uppermost in his thoughts. We still cannot argue that this is conclusive proof that Wagner read Göttling while there is no external evidence to sustain it: Wagner's inventive and imaginative powers were easily the equal of those of any Romantic scholar, and we may not rule out the possibility that the parallel creation was the result of spontaneous inspiration. Yet this new evidence of exclusive influence from Göttling's *Nibelungen und Gibelinen*, coupled with the other instances we saw earlier in the chapter, makes it difficult to believe that the correspondence is coincidental, and together the evidence forms a reasonable case for arguing that Göttling's Nibelung study influenced Wagner's 1848 creation.

[73] 'Brunhild . . . folgte still dem Gunnar und ward sein Weib in Burgund, und Sigurd erkannte sie immer nicht.' (*Nibelungen und Gibelinen*, p. 49.)

[74] 'Mit dem strömenden Herzblut die Erinnerung an seine verlorne Seligkeit mit Brunhilden wiederkehrte.' (Ibid.)

7

Young Siegfried

AFTER a break of almost two and a half years Wagner, now in Zurich, resumed work on his Nibelung drama in early May 1851 with *Der junge Siegfried*, forerunner of *Siegfried*. In the interim we can assume he had more or less completed his *Ring* reading, including works with such bearing on the Young Siegfried drama as the *Thidreks saga*, if he had not already read it before; the *Völsunga saga*; and almost certainly, in the two months preceding the new drama, Simrock's translation of the Edda.

The *Völsunga saga* and *Thidreks saga* completed the spectrum of Young Siegfried stories and, since Wagner had rejected the *Nibelungenlied* version (see Chapter 1), marked the opposite ends of it. The *Völsunga saga* Siegfried, following the same tradition as the Eddas, grows up at the court of his royal stepfather under the tutelage of the smith Reigen. Before he will fight the dragon he exacts a sword from his foster-father and wins a glorious and gory victory over his father's slayers.

The Siegfried of the *Thidreks saga*, akin to his counterpart in *Das Lied vom Hürnen Seyfrid*, is a true comic-strip hero. Endowed with enormous but sadly ill-directed strength, Siegfried beats the smith's apprentices and when he tries to 'help' practically wrecks the smithy. The smith remonstrates, whereupon Siegfried slinks off and holds his tongue. In the second instalment Siegfried sets off early one morning to burn charcoal. By mid-morning he has stripped part of the forest and demolished nine days' provisions, and when the dragon appears he cudgels him with a tree-trunk.

In his autobiography Wagner refers to his Young Siegfried drama as a 'heroic comedy'.[1] This diplomatic term both

---

[1] 'heroisches Lustspiel' (*ML*, p. 478).

acknowledges the full range of the Siegfried tradition and indicates Wagner's intention to embrace both ends of it, comic and heroic, in his drama. The source material was abundant; yet for all that, finding middle ground between such disparate sources, between the high heroics of the *Völsunga saga* and the knockabout comedy of the *Thidreks saga*, was not an easy task. Not surprisingly, the course Wagner adopted shows his indebtedness to the pioneering work of two predecessors, Fouqué and Simrock.

Fouqué had been the first to work out a plausible combination of the comic and heroic traditions. Both he and Simrock, whose version in the *Amelungenlied* runs along very similar lines, favoured the court upbringing described in the *Völsunga saga* more than did Wagner. Both, however, were aware of the alternative version of Siegfried's youth and both, like Wagner, thought it too important to be merely ignored.

The solution Fouqué thought out and Simrock later followed involved retaining the physical quality of the action *à la Thidreks saga* but transforming it inwardly. Siegfried's antics in *Das Lied vom Hürnen Seyfrid* and the *Thidreks saga* are burlesque; but when he makes to grasp the red-hot sword-hilt or forces the smith to run for his life in Fouqué, or when, more traditionally, in Simrock the apprentices and the anvil take a hammering, it is conceived as part of Siegfried's heroic temperament: part imperiousness, part the hastiness of a fiery disposition, and above all impatience and frustration at the delay in the delivery of a suitable sword. So Simrock's Siegfried, having broken both of the master smith's swords against the anvil, resolves to give a demonstration lesson:

> Aller Hämmer schwersten nahm er in die Hand.
> 'Achtung, dass ihr was lernet,' rief er zornentbrannt.
> Oa schlug er auf die Stange einen Schlag, er war nicht krank.
> Der Stein zerbarst, der Amboss in der Erde Grund versank.
>
> .   .   .   .
>
> 'So sollt ihr mir schmieden,' sprach Siegfried, 'fortan;
> Morgen komm ich wieder, und wer es da nicht kann,
> Den schweiss ich auf den Amboss . . .'[2]

[2] *Heldenbuch*, iv. 94.

The heaviest of all hammers he took up in his hand.
'Watch out and learn from this,' he cried out, fuming, and
He hit upon the bar then a blow: it was quite sound—
The stone split, the anvil sank into the ground.

   .   .   .   .

'That's how you should hammer,' said Siegfried, 'from now
   on.
I'll be back tomorrow; if anyone can't do it then,
I'll weld him to the anvil . . .'

Or, as Fouqué's Sigurd explains breathlessly to his mother
when his pursuit of the smith is interrupted:

> Der Reigen—O das Alles ist so lang—
> Er schmiedet, schmiedet,—lobt sein eignes Werk,
> Und klirr! dann bricht's bei meinem ersten Hieb,—
> Und ohne Waffen ich—lass mich ihn fassen![3]

> That Reigen—oh, it's all so long—
> He hammers, hammers, praises his own work,
> And clink! at my first blow it breaks
> And I am weaponless—let me catch him!

Between them Fouqué and Simrock had shaped a new
Siegfried tradition, that of the overbearing and impetuous
youth, given to direct physical intervention in furtherance of
his wishes. In Wagner he is the 'hasty boy'[4] who grabs the
smith by the throat, knocks the pot out of Mime's hands and
brings along a bear to hurry him up.[5]
But Fouqué's reworking of the Siegfried tradition went
both ways. If on the one hand he borrowed some of the
sturdy physical approach of the *Thidreks saga* and *Das Lied
vom Hürnen Seyfrid* for his 'heroic' Siegfried, on the other he
lent articulation to the dumb lad of the 'comic' version.
To Fouqué must go the credit for having devised a dimension of verbal aggression to match the physical aggression of
the sources. Simrock was not far behind. As Wolfgang Golther notes, the influence of these two writers on the language
and dialogue patterns of the first act of *Siegfried* is quite

---

[3] *Sigurd*, p. 19.
[4] 'hastiger Knabe' (*Schr.*, vi. 90).
[5] *Schr.*, vi. 95, 89, 87.

transparent.[6] Golther quotes from the sword-testing in Fouqué's *Sigurd*:

> *Sigurd*
> *(Er haut gegen den Eckstein. Die Klinge zerspringt.)*
> Sieh den vermaledeiten Binsenstock!
>   *Reigen*
> Das? Binsenstock?
>   *Sigurd*
> Ja, Hält's denn besser vor? [. . .]
> Seht mir den Prahler, seht den trägen Werkmann!
> Willst du nicht tüchtig schmieden? So thu ich's,
> Und zwar auf deinen Kopf an Amboss statt,
> Dazu noch ist des Schwertes Trümmer gut.[7]

> *Sigurd*
> *(He strikes a blow against the corner-stone. The blade*
>   *shatters.)*
> Look at the wretched sedge-stalk!
>   *Reigen*
> A sedge-stalk? That?
>   *Sigurd*
> Does it bear up better? [. . .]
> Just look at the boaster! Look at the idle workman!
> If you won't hammer properly, then I will,
> And namely on your head in place of anvil:
> The broken sword's still good enough for that.

He could equally well have quoted from Simrock, who uses similar menacing language:

> 'Das ist nun dein Geschmiede,' sprach da Siegfried,
> 'Mime, greiser Prahlhans, du unnützer Schmied,
> Kannst du nichts Bessres wirken, als solch ein gläsern
>   Ding,
> So bist du zum Erschlagen, zum Hängen selbst zu gering.
>   .   .   .   .
> 'Ich hätte Lust und würfe dir ins Gesicht das Heft.'
> 'Dir schmieden,' sprach da Mime, 'das ist ein übles
>   Geschäft.'[8]

> 'That's your handwork for you,' then said Siegfried;
> 'Mime, you hoary braggart, you good-for-nothing smith,

---

[6] Golther, *Grundlagen*, pp. 64–5.
[7] *Grundlagen*, p. 65.
[8] *Heldenbuch*, iv. 93.

If such a glass-like thing's the best you can do,
You're not worth the trouble of killing, even hanging's too
   good for you.

.   .   .   .

'I rather feel like throwing the sword hilt in your face.'
'It's a bad business forging for you,' said Mime at this.

Wagner embraced his predecessors' strong mode of expression with evident enthusiasm. When Mime's sword fails the test Siegfried bursts out with all the abusive and violent language we have just witnessed, along with the signs of impatience and frustration found in the earlier passages we quoted:

> Hei! was ist das
> für müss'ger Tand!
> Den schwachen Stift
> nennst du ein Schwert?
> *(Er zerschlägt es auf dem*
> *Amboss, dass die Stücken*
> *ringsum fliegen: Mime weicht*
>   *erschrocken aus.)*
> Da hast du die Stücken,
> schändlicher Stümper;
> hätt' ich am Schädel
> dir sie zerschlagen! —
> Soll mir der Prahler
> länger noch prellen?
> Schwätzt mir von Riesen
> und rüstigen Kämpfen,
> von kühnen Taten
> und tüchtiger Wehr;
> will Waffen mir schmieden,
> Schwerte schaffen;
> rühmt seine Kunst,
> als könnt' er was Rechtes:
> nehm' ich zur Hand nun,
> was er gehämmert,
> mit einem Griff
> zergreif' ich den Quark! —
> Wär' mir nicht schier
> zu schäbig der Wicht,
> ich zerschmiedet' ihn selbst

mit seinem Geschmeid' . . .[9]
Hey! What kind of
trinket's this!
You call that slender
pin a sword?
*(He shatters it on the anvil in
such a way that fragments fly
all round. Mime, alarmed,*
takes avoiding action.)
There's the pieces,
blasted bungler;
I should have smashed them
on your skull!
Shall I let the coxcomb
con me again?
Blabs of giants
and action in battle,
of bold deeds
and doughty weapons;
says he'll forge me arms,
fashion me swords,
blows his own trumpet,
claims skill at his trade—
I take in my hand then
what he's been hammering:
with one movement
I've mangled the trash!
If the little manikin
weren't just too mean,
I'd smash smith and handwork
to smithereens . . .

Other examples could be quoted. But the effective combination of the two Siegfried traditions did not exhaust the influence of these two Nibelung writers on the Young Siegfried drama; both Fouqué and Simrock made further, differing contributions.

It was Fouqué, for instance, who suggested the scene structure for *Siegfried* with his prologue in the smithy, the first Adventure on Gnitaheide, and the second at Brünn-

[9] *Schr.*, vi. 88–9.

hilde's rock. Fouqué had also eliminated Siegfried's revenge expedition from the stage action, relegating it to retrospective narration in the Gnitaheide scene.[10]

In addition, Fouqué was responsible for the deterioration of the relations between smith and foster-son *via-à-vis* the sources and indeed Wagner's own relatively harmonious account in the *Mythus* (see Chapter 5). Possibly Fouqué felt that since the sources later oblige Siegfried to kill the smith it would be better if they started out on the wrong footing. In any case, Fouqué takes the somewhat fractious, sour note that marks their dealings in the *Völsunga saga* and turns it into a full-blooded antipathy, a mutually damaging incompatibility between different types. In Wagner Siegfried's revulsion may border on the neurotic:

> Seh' ich dir erst
> mit den Augen zu,
> zu übel erkenn' ich,
> was alles du tust:
> seh' ich dich stehn,
> gangeln und gehn,
> knicken und nicken,
> mit den Augen zwicken:
> beim Genick möcht' ich
> den Nicker packen,
> den Garaus geben
> dem garst'gen Zwicker![11]

> I only need
> to use my eyes
> to see evil
> in all you do:
> when I see you standing,

---

[10] Wagner had included the revenge expedition in the *Mythus* (*Skizzen*, p. 28) and in *Siegfrieds Tod* (*Schr.*, ii. 219). When he came to his stage version of the Young Siegfried drama he decided to dispense with the revenge by making Siegfried ignorant of his father's slayer (and later, in *Die Walküre*, by arranging for Hunding's immediate death). One trace, however, still remains in Act III of *Siegfried*: when Siegfried wields his sword against Wotan's spear it is partly because he deduces, not totally incorrectly, from what the Wanderer has told him that the god is responsible for his father's death: 'Meines Vaters Feind! | Find' ich dich hier? | Herrlich zur Rache | geriet mir das!' ('My father's enemy! | Have I now found you? | What an invaluable | chance for revenge!' *Schr.*, vi. 163.)

[11] *Schr.*, vi. 91.

shuffling and shambling,
bobbing and nodding,
squint eyes mopping—
I could pack old Noddy
by the neck,
finish him off,
the foul twitcher.

Yet it is only a step away from the sentiment Fouqué's Sigurd
expresses just before putting a final end to the bickering:

Das ist ja auf die Art ein ganz verworfner,
Verruchter Bursch, und Allem, was die Welt
Rechtliches trägt und Schönes, thäte man
Den Besten Dienst, wenn man solch Ungethüm
Abschlachtete, vor Schaden Andre hütend.[12]

In his manner he's a quite depraved
And rascally fellow, and to all that's just
And lovely in the world the greatest service
One could do would be exterminating
Such a monster, keeping folk from harm.

One final matter in which Fouqué helped Wagner was the
disposal of the accumulated corpses after the dragon-fight. At
the same time he solved the problem of what Siegfried
should do with the hoard he has inadvertently won. In the
sources Siegfried takes the gold after the dragon fight, even if
he does not always get very far with it;[13] Wagner wanted to
show Siegfried immune to the lure of the gold. Fouqué's
Sigurd does indeed eventually load the gold on Grani, faith-
ful to the *Völsunga saga* and the Eddas, and ride away with it,
but before doing so he muses on the hoard and the curse he
has seen fulfilled and reflects:

Wär vielleicht
Wohl klug gethan, die beiden hässlichen
Blutrothen Brüder hier sammt ihrem Schatz
In Gnitahaides Dunkelheit zu lassen.[14]

---

[12] *Sigurd*, p. 54.
[13] In *HS* Siegfried soon sinks it in the Rhine.
[14] *Sigurd*, p. 59.

It would perhaps
Be no bad thing to leave in Gnitaheide's
Darkness both the hideous blood-stained brothers
Lying here, together with their hoard.

Wagner's Siegfried follows the wise counsel of Fouqué's Sigurd: he leaves the hoard in the cave, drags Mime's body on to the gold, rolls Fafner's up against the entrance, and declares:

Da lieg' auch du,
dunkler Wurm!
Den gleissenden Hort
hüte zugleich
mit dem beuterührigen Feind.[15]

You lie here too,
dark dragon!
The glittering hoard
guard together
with the enemy eager for spoil.

If Fouqué's Sigurd is bumptiously heroic,[16] the influence of Simrock's Siegfried lies in the other direction. On the day he completed the sketch for *Der junge Siegfried* Wagner wrote to Ferdinand von Ziegesar of the 'comic, well-nigh popular material for a Young Siegfried drama'.[17] The idea, he explained, was that the public should get to know the story of Young Siegfried 'in its most popular aspects . . . effortlessly, so to speak, like children getting to know it in a fairy-tale'.[18] It was to this lighter mood, the infectious spirit, the popular and folk strain in Wagner's drama that Simrock made his most indelible contribution.

Simrock's influence on the poetic mood of Act II has already attracted comment from Golther.[19] Describing Siegfried as he sallies forth into the forest early on the morning of his dragon-fight, Simrock evokes in verse the atmosphere of

---

[15] *Schr.*, vi. 148–9.
[16] 'Zu Pferd ist adlich kecker Fürsten Sitz' ('A bold and high-born prince belongs on horseback') is just one example of Sigurd's priceless utterances: *Sigurd*, p. 41.
[17] 'heitere und fast populäre Stoff zu einem jungen Siegfried' (letter, 10 May 1851, *Dokumente*, p. 43).
[18] 'in den populärsten Zügen . . . gewissermassen spielend, wie ihn Kinder durch ein Märchen kennen lernen'
[19] *Grundlagen*, pp. 69–70.

the woodland setting which Wagner was to portray so magi-
cally in the 'Waldweben' music:

> Noch stand die Sonne niedrig, da fuhr zum grünen Wald
> Siegfried der junge; wie fröhlich ward er bald
> Als er im lichten Scheine die Baume grünen sah;
> Vor freuden wollt er springen, nicht wusst er, wie ihm
>   geschah.
>
> Er begann ein Lied zu singen: nach sangs der Wiederhall:
> Da schuf ein lustig Ringen der starken Stimme Schall.
> Bald freut' ihn mehr zu lauschen des Bächleins muntrem
>   Gang,
> Bald wie ein wonnig Rauschen durch alle Läuber sich
>   schwang.
>
> Von abertausend Stimmen der Wald erfüllet war,
> Von Blüthen summten Immen zu Blüthen immerdar,
> Bald Adlersflügelschläge, bald kleiner Vögel Lied,
> Bald Reh im Laube raschelnd, bald Wasservögel im Ried.[20]
>
> The sun was still low in the sky when Siegfried the boy
> Went into the greenwood; soon he was full of joy
> As the clear light showed him the new leaves on the tree;
> He could have jumped for joy then, so amazed was he.
>
> The boy struck up a song: back the echo sang;
> The lusty voice's sound in merry contest rang.
> Now the blithely flowing stream chiefly caught his ear,
> Now a delightful rustling, setting leaves astir.
>
> A thousand thousand voices crowded through the trees:
> Ever on from flower to flower hummed the honey-bees;
> Now the beat of eagle's wing, the song of tiny bird,
> The deer now crackling through the leaves, the marsh fowl
>   now he heard.

Tales of other heroes in Simrock's *Amelungenlied* also left
their trace on Wagner's drama. Siegfried's sword-forging,
where he horrifies master Mime by filing Notung to swarf,
suggests that Wieland was his tutor:

> Wieland in der Schmiede nahm eine Feile gut;
> Damit war zerfeilet das Schwert zu eitel Staub . . .[21]

[20] *Heldenbuch*, iv. 99.
[21] *Heldenbuch*, iv. 57. Wieland's method, 3 times repeated, results in his
masterpiece Mimung. Wagner could of course also have known the episode direct
from the *Thss*, from which Simrock took it.

Wieland in the smithy took a file in hand,
And with it filed the sword down to specks of dust . . .

Simrock's 'Dietleib' gave rise to a little comic incident in Act II. Dietleib in Simrock's version is a hero of the same school as the 'comic' Siegfried. A late developer, the despair of his parents, Dietleib spends an unpromising boyhood lounging around the kitchens before maturing into a doughty hero. On his first independent adventure he comes to Sintram's court and spies a horn waiting to be blown:

> Ein Horn lag auf dem Stuhle. Verstand ers Blasen auch?
> Er nahm es von dem Polster und hob es an den Mund:
> Da fing er an zu blasen ob er es gleich nicht verstund.
>
> Er mocht es selber merken, dass ihm die Kunst noch
>   fremd;
> Doch wollt er sie erlernen . . .[22]
>
> On the chair a horn lay. Could he play it too?
> He took it off the cushion and put it to his mouth,
> Then he began to play it, although he didn't know how.
>
> He saw it was an art-form he hadn't mastered yet,
> But he wished to learn it . . .

Nothing daunted, Dietleib carries on lustily as though determined to rouse the beasts of the forest:

> Er aber blies, als wollt er Eber, Bär und Wolf
> Aus dem Walde blasen . . .[23]
>
> He blew as though he wanted to call up from the wood
> Boar and bear and wolf . . .

In the end it is Sintram who appears and challenges him to combat.

Out of this Wagner developed the débâcle with the reed-pipe. Siegfried fashions the pipe to talk to the wood bird, but cannot get the knack of playing it:

> Das tönt nicht recht;
>   auf dem Rohre taugt
> die wonnige Weise nicht:—
>   Vöglein, mich dünkt,

---

[22] *Heldenbuch* v. 37.
[23] Ibid.

> ich bleibe dumm:
> von dir lern' ich nicht leicht![24]

> That doesn't sound right:
> the ravishing tune
> can't be played on a pipe:—
> Wood-bird, I fancy
> I shall stay foolish:
> I can't learn quickly from you!

Now Siegfried reaches for the horn with which he lures the forest animals and plays his own tune:

> Nach lieben Gesellen
> lockt' ich mit ihr:
> nichts bessres kam noch
> als Wolf und Bär.[25]

> Delightful comrades
> I've coaxed with this:
> none better came yet
> than wolf and bear.

This time it is Fafner who answers and battle is joined.

Simrock also dipped into fairy-tales in the course of his Siegfried story. Brünnhilde's awakening is based closely on Sleeping Beauty, complete with the kiss which the Scandinavian sources omit but Wagner includes.[26] It is possible that Simrock also had some influence on the astonished Mime's exclamation when Siegfried's revolutionary sword-forging looks like succeeding:

> Nun ward ich so alt
> wie Höhl' und Wald,
> und hab nicht so 'was gesehn![27]

> Now I am as old
> as cave and wood
> and I've never seen the like!

As has been pointed out,[28] the little motif echoes the change-

---

[24] *Schr.*, vi. 135.

[25] Ibid.

[26] The episode is narrated by Heime in 'Wittich Wielands Sohn', Abenteuer 20 (Simrock, *Heldenbuch*, iv. 368–75).

[27] *Schr.*, vi. 115.

[28] e.g. Golther, *Grundlagen*, p. 75.

ling's give-away cry in a tale from Grimms' *Märchen*, 'Die
Wichtelmänner' ('The Elves'):

> nun bin ich so alt
> wie der Westerwald,
> und hab nicht gesehen, dass jemand in Schalen kocht![29]

> now I am as old
> as the Westerwald,
> and I've never seen anyone cooking in eggshells!

In the *Thidreks saga*, Simrock's source for his scene, Mime
also registers surprise at Siegfried's activities at the anvil: 'I
never saw a more fearsome nor a clumsier blow from
anyone.'[30] Simrock combined the anvil scene with the fairy-
tale motif in a way which may have persuaded Wagner to do
likewise; in the resulting scene Mime, watching Siegfried
deliver a massive blow to the anvil, calls out:

> Nun leb ich siebzig Jahre und drüber manchen Tag,
> Und nimmer sah ich, nimmer einen fürchterlichern Schlag,
> Als den auf diese Stange ein Kind hat geführt.[31]

> I have been around now for seventy years and more
> And a blow more fearsome than this I never saw,
> Which a child has dealt upon this iron bar.

The influence of the odds and ends of fairy-tale Wagner
found in Simrock was, however, at most only marginal. His
real mentors in fairy-tale matters were the Grimm brothers.
It was in their collection that Wagner had found the 'Wichtelmänner' tale; they had also shown him, quite independently of Simrock, the association between Brünnhilde
and Sleeping Beauty: 'The maiden sleeping in the castle surrounded by a barrier of thorns until the chosen prince
delivers her is the sleeping Brunhild . . . The spindle . . . is
the "sleep thorn" which Odin pricks Brunhild with.'[32]

---

[29] i. 205. Grimm also draws attention to these lines in the ch. 'Wichte und Elbe' of
his *Mythologie*, pp. 437–8.
[30] 'Niemalen sah ich von jemandem einen fürchterlicheren noch ungefügeren
Schlag.' (ii. 28.)
[31] *Heldenbuch*, iv. 94.
[32] 'Die Jungfrau, die in dem mit einem Dornenwall umgegebenen Schloss schläft,
bis sie der rechte Königssohn erlöst, ist die schlafende Brunhild . . . Die Spindel . . .
ist der Schlafdorn, womit Othin die Brunhild sticht.' (*Märchen*, iii. 87.) See also J.
Grimm, *Siegfrieds Tod*, p. 390 and W. Grimm, *Heldensage*, p. 384.

Most pertinent of all were the Siegfried tales which the Grimm brothers claimed to have discovered.[33] These are, in the main, tales of the simple lad who, aided by luck and resourcefulness, overcomes all obstacles, releases the enchantment, wins the hoard, and marries the princess. The tale which the Grimm brothers single out as most typifying Siegfried is the first of the series, 'Der junge Riese' ('The Young Giant'). In the introduction to their *Märchen* we read: 'Siegfried, for instance, frequently appears at his most recognizable in the Young Giant, characterized by that unique blend of a pure, brave heart and a good-natured, playful humour'.[34]

The Young Giant displays an unmistakable affinity with the Siegfried of the *Thidreks saga* and *Das Lied vom Hürnen Seyfrid*. After eating his parents out of house and home the Young Giant sets forth into the world and becomes apprenticed to a smith. His first almighty blow sinks the anvil deep into the ground. For wages he satisfies himself by dealing buffets and kicks.

The Grimm brothers taught Wagner to see Siegfried in latter-day fairy-tales: Wagner looked further, and came up with a thrilling discovery which he wrote about to Theodor Uhlig with the adrenalin still high from his work on the Young Siegfried sketch:

Didn't I write to you once before about a comic subject? It was about the boy who sets out 'to learn how to fear' and is so simple that he can never manage to learn it. Just think of the thrill I got when I suddenly realized that this boy is none other than—young Siegfried, who wins the hoard and wakes Brünnhilde![35]

The lad in the fairy-tale, No. 4 in the Grimm brothers' collection, is a layabout totally lacking in ambition until one

---

[33] See ch. 1 n. 33.

[34] 'namentlich erscheint Siegfried öfter am kenntlichsten in dem jungen Riesen', an jener eigenthümlichen Mischung eines tapfern und reinen Herzens und einer gutmüthigen und scherzhaften Laune' (i. p. xlviii).

[35] 'Habe ich Dir nicht früher schon einmal von einem heitern Stoffe geschrieben? Es war dies der Bursche, der auszieht "um das Fürchten zu lernen" und so dumm ist, es nie lernen zu wollen. Denke Dir meinen Schreck, als ich plötzlich erkenne, dass dieser Bursche niemand anders ist, als—der junge Siegfried, der den Hort gewinnt und Brünnhilde erweckt!' (Letter, 10 May 1851, *Dokumente*, p. 42.)

day, fascinated by his brother's account of the sensations that beset him when he passes by the churchyard at night, he resolves to go out into the world and learn for himself how to tremble. In the course of a series of spirited and ghoulish adventures, including a variant of the by now familiar smithy scene, the boy wins the hoard, releases the castle from enchantment, and marries the princess. Alas, he still has not learnt to tremble. Finally his wife comes to the rescue: as he lies asleep one night she douses him with teeming water from the brook. He wakes up trembling and they live happily ever after.

Translated into the terms of the Siegfried saga, Wagner's version of the 'learning to fear' tale reads as follows: Siegfried knows fear only at second-hand from Mime, who is all too familiar wih the sensation. In order to learn it for himself Siegfried agrees to fight Fafner: but Fafner cannot help. His quest takes him through the fire surrounding Brünnhilde's rock, but he is no wiser than before. Only a woman can teach Siegfried to tremble. On the other side of the flame barrier he finds Brünnhilde asleep, and is overcome by the most violent sensations:

> Brennender Zauber
> zückt mir ins Herz;
> feurige Angst
> fasst meine Augen:
> mir schwankt und schwindelt der Sinn!
>
> .   .   .   .   .   .
>
> Wie ist mir Feigem?
> Ist es das Fürchten?
> O Mutter! Mutter!
> Dein mutiges Kind!
> Im Schlafe liegt eine Frau:—
> die hat ihn das Fürchten gelehrt![36]

> Burning enchantment
> bursts through my heart;
> fiery terror
> holds my eyes fast:
> my senses stagger and sway!

---

[36] *Schr.*, vi. 165.

.   .   .   .

What is this folly?
Is this what fear is?
Oh mother, mother!
Your manly son!
A lady lies fast asleep:—
from her he has learnt how to fear!

The connection between the fairy-tale boy and Siegfried was
Wagner's own personal 'find'. In some ways it comes as a
surprise that the Grimm brothers, whose eyes were constantly open to such associations, did not recognize the pos-
sibilities of the boy who could not learn fear, for the Siegfried
sources themselves are full of indicators. According to the
Thidreks saga, the *Völsunga saga* and the Edda fearlessness is
Siegfried's hallmark. In the practically identical descriptions
of Siegfried in the *Thidreks saga* and *Völsunga saga* we read:
'his courage never failed, and never in his life was he afraid.'[37]
'He never lacked courage and he was never afraid.'[38] In the
Eddic poem *Helreidh Brynhildar* the disobedient valkyrie describes how Odin eliminated other possible awakeners:

> *Dem* gebot er
> Meinen Schlaf zu brechen,
> Der immer furchtlos
> Würd erfunden.[39]

> Him he bade
> Break my sleep
> Who was always found
> To be without fear.

Siegfried hears the same thing from the valkyrie in
Sigrdrífumál.[40]

Before his Young Siegfried inspiration struck, Wagner
told Uhlig, the tale of the boy who sets out to learn fear had

---

[37] 'niemalen entstand ihm der Muth, und niemalen in seinem Leben ward er
erschrocken.' (*Thss*, ii. 70.)
[38] 'Nie ermangelte er des Muthes, und niemalen ward er erschrocken.' (*Vs*,
p. 109.)
[39] Simrock, *Edda*, p. 188.
[40] Simrock, *Edda*, p. 169.

been plaguing him all winter.[41] We can imagine Wagner in
the spring of 1851, his head humming with the 'learning to
fear' story, reading Simrock's newly published *Edda*; as he
reached the above lines in the *Sigrdrífumál* or *Helreidh Brynhildar* he must have experienced a shock of recognition
equivalent to the one he invented for Siegfried at the close of
Act II:

> Der dumme Knab',
> der das Fürchten nicht kennt, —
> —das ist ja der Siegfried![42]

> The foolish boy
> who doesn't know fear—
> —of course, that's Siegfried!

The formulation of his 'heroic-comic' Young Siegfried was
one of two profound developments in Wagner's new drama
as against the work of 1848. The other came on the mythological side. The deepening of the mythology is particularly
noticeable by comparison with Wagner's earlier Siegfried
drama. In *Siegfrieds Tod* the gods do not appear; they are cult
figures, to be invoked in rituals and ceremonies. Their function is to watch over the actions of men and their indi-
viduality is confined to role epithets—'Preserver of
Oaths', 'Guardian of Marriage'—and various accoutrements:
Wotan's ravens, Fricka's sheep, Thor's thunderbolts. Anti-
quarian and essentially 'operatic', Wagner's handling of
ancient Germanic religion in *Siegfrieds Tod* rather harks back
to Felice Romani's libretto for Bellini's *Norma* than points
forward to the direction he was later to take.

In the Young Siegfried drama cult religion is left behind
and living myth in the formation takes its place. For the first
time one of the gods walks the stage in the guise of the
'Wanderer', whose four scenes punctuate the drama. There
had been no intimation of the Wanderer scenes in the *Mythus*,
and their appearance in *Der junge Siegfried* was doubtless

---

[41] Letter, 10 May 1851, *Dokumente*, p. 42.
[42] Wagner's original 3rd line runs: 'mein Vöglein, der bin ja ich!' ('Woodbird, of
course, that's me!': *Schr.*, vi. 151).

the fruit of Wagner's post-*Mythus* reading of the *Völsunga saga*.

Although Odin manifests himself in different disguises in the Eddas too, it is the *Völsunga saga* which really creates the figure of the mysterious stranger who enters the scene at critical junctures in the Wälsung family history: at Signy's wedding and Sigmund's death, as Siegfried chooses Grani, on his revenge voyage, and at Gnitaheide before the dragon fight. Typically he appears as an old man with a beard, wearing a blue cloak and a hat pulled low over his face, perhaps to conceal a missing eye. In case there should be any doubt as to the stranger's true identity, von der Hagen obligingly puts in the footnotes of his translation that this is Odin.[43] Wagner correspondingly describes the Wanderer on his first appearance at the entrance of Mime's cave: 'He is of lofty appearance and one-eyed; brown, curly beard; long, dark-blue cloak; he carries a spear for a staff. On his head he has a large hat with a broad, round brim which hangs down low over the missing eye.'[44]

The *Völsunga saga* supplied the description of the Wanderer and his presence in the Siegfried saga. The name and much of the material of Wagner's four Wanderer scenes came from the Edda. Three of the Eddic 'Götterlieder' tell of journeys of observation and enquiry which Odin undertakes. Each time he travels under an assumed name: Grimnir on his visit to King Geirröd, Gangradr in his dispute with the giant Vafthrudnir, Vegtam when he questions the vala. Odin's assumed names had been glossed variously by different translators in the past; Simrock in his new translation took Odin's names in the three poems and came up with a common identity, a new standard image for the travelling god: that of the 'Wanderer'. In his notes to *Vafthrudnismál* Simrock writes of the name Gangradr: 'Like Gangleri, which according to *Grimnismál* [verse] 46 is also one of Odin's names . . .

[43] *Vs*, pp. 13, 54, 63, 77, 82. Odin's appearance during Siegfried's revenge voyage is something Wagner would have already known from his reading of the Eddic 'Heldenlieder'.

[44] 'Er ist von hoher gestalt, und einäugig; brauner, lockiger bart; dunkelblauer langer mantel: er trägt einen speer als stab. Auf dem haupte hat er einen grossen hut mit breiter runder krämpe, die über das fehlende auge tief hereinhängt.' (*Skizzen*, p. 118.) The description in the present *Siegfried* libretto is similar but less full.

like Vegtam, which Odin assumes in the *Vegtamskvidha*, this name signifies "the Wanderer".'[45]

The name 'Wanderer' is just one of many indications that Wagner had been reading Simrock's new complete Edda translation in the couple of months or so between its publication and his first sketch for the Young Siegfried drama. Certainly Wagner's interest in the Eddas, particularly the *Poetic Edda*, had been significantly revitalized. The Eddic presence in *Der junge Siegfried* is remarkably fresh, vigorous, and intense, no longer, as in the work of 1848, concerned almost exclusively with the plot of the drama but now directed more towards the texture. Particularly in the Wanderer scenes the spirit of the Eddas breathes through the form and style, the mood, language, and idiom. It was perhaps Wagner's power to create afresh in Edda terms as much as his provision of a unified plot which inspired Heinrich von Stein to comment:

Actually, it is quite simply as though the *Ring* were an original text which was no longer available to the Edda poets and partial comprehension of which can therefore be traced only here and there in their poems: so much does everything in the drama seem to be restored to its original unity and to be newly created and brought to life.[46]

Wagner's new Eddic mode can be seen at its best in two of the Wanderer scenes, the Act I knowledge contest with Mime and the summoning of the vala in Act III. Each has a structural basis in one of the Wanderer poems of the Edda. The influence of *Vafthrudhnismál* on Wotan's encounter with Mime is widely recognized.[47] In the Eddic poem Odin, disguised as Gangradr, appears travel-weary in the doorway of

---

[45] 'Dieser Name bedeutet wie *Gangleri*, der nach Grimnism. 46 gleichfalls einer von Odins Namen ist . . . wie *Wegtam*, den Odin in der Wegtamskvidha annimmt, den Wanderer.' (*Edda*, p.346.) In the other edns. of the Edda Wagner knew, just once is one of Odin's names, Gangrád, glossed as 'Wanderer', by Studach (*Sämund's Edda des Weisen*, p. 64.)

[46] 'Es ist schliesslich einfach und geradezu, als ob der Ring eine den Eddadichtern nicht mehr zugängliche gewesene Urschrift sei, deren teilweises Verständniss man demnach in ihren Liedern nur hie und da verspürte: so sehr scheint im Drama alles zu seiner Ureinheit zurückgeführt und neu geschaffen und belebt.' (*Bayreuther Blätter* (1899), p. 189; quoted in Golther, *Grundlagen*, p. 7.)

[47] See e.g. Cooke *World*, p. 111; Golther, *Grundlagen*, pp. 77–9.

the wise giant Vafthrudnir and challenges him to a contest of knowledge. Odin answers the giant's questions satisfactorily and takes his seat on the bench. Next Vafthrudnir stakes his own head on being able to answer his guest:

> Das Haupt zur Wette hier
> Steh in der Halle,
> Gast, um weise Worte.[48]

> My head stand forfeit
> Here in the hall,
> Guest, for wise words.

After their discourse has ranged through the universe from creation to destruction and rebirth, Odin eventually traps the giant by what can only be termed a trick question:

> Was sagte Odhin
> Ins Ohr dem Sohn,
> Eh er die Scheitern bestieg?[49]

> What did Odin say
> In his son's ear
> Before he mounted the pyre?

Vafthrudnir recognizes the identity of his opponent, and with it defeat:

> Nicht Einer weiss
> Was in der Urzeit du
> Sagtest dem Sohn ins Ohr.[50]

> No one can tell
> What in ancient times
> You said in the ear of your son.

Obviously Wagner used the framework of *Vafthrudhnismál* in the knowledge-contest scene between Wotan and Mime in Act I of his Young Siegfried drama. The weary traveller, the knowledge of things far and near, with hospitality the reward and death the price of failure: all recur in Wagner's poem. Some of the content is there too, since Wagner uses Mime's questions and Wotan's answers to sketch in his cosmos. Even the language has taken root: the Wanderer's

---

[48] Simrock, *Edda*, p. 22.
[49] Simrock, *Edda*, p. 26.
[50] Ibid.

Viel erforscht' ich,
erkannte viel . . .[51]

Much I sought,
discovered much . . .

harks back to Gangradr's

Viel erfuhr ich,
Viel versucht ich,
Befrug der Wesen viel.[52]

Much I learnt,
Much I tried,
Questioned many beings.

The influence of *Vafthrudhnismál* was powerful. But another Eddic poem also helped shape this scene: *Alvíssmál*.[53] Simrock describes *Alvíssmál* as an imitation *Vafthrudhnismál*, but this time with the emphasis on comedy in the framework.[54] The dwarf Alvíss—'Know-all'—comes to demand the hand of Thor's daughter in marriage. Thor consents— provided the dwarf can answer all his questions. The dwarf is so absorbed in displaying his knowledge that he forgets what no dwarf can afford to forget: to be safely underground by daybreak. Thor keeps Alvíss talking until sunrise and the dwarf is doomed.

The foolish vanity of the dwarf Alvíss, whose ideas range far and wide but who neglects the things nearest home, was obviously a model for Mime. As with the *Vafthrudhnismál*, some of the language has rubbed off on to Wagner's scene. In the Eddic poem Thor refers to the dwarf ironically in every question as 'clever dwarf', while the Wanderer, after Mime has answered the first two questions correctly, treats the dwarf to:

Der witzigste bist du
unter den Weisen:
wer käm' dir an Klugheit gleich?[55]

[51] *Schr.*, vi. 100.
[52] Simrock, *Edda*, p. 25.
[53] Cooke, *World*, p. 111.
[54] *Edda*, p. 345.
[55] *Schr.*, vi. 107.

> Among all the wise
> you've the sharpest wits:
> who could compare for wisdom?

His next question completely stumps Mime.

The Wanderer's summoning of the vala in Act III, it is commonly agreed,[56] owes much to the Eddie *Vegtamskvidha*. Travelling this time under the name Vegtam, Odin seeks out the vala in her grave-mound at the doors of hell. Unwillingly she rises to Odin's incantation:

> Welcher der Männer,
> Mir unbewuster,
> Schafft mir Beschwer,
> Stört mir die Ruh?
> Schnee beschneite mich,
> Regen beschlug mich,
> Thau beträufte mich,
> Todt war ich lange.[57]

> Who is the man
> Unknown to me
> Who brings me trouble,
> Disturbs my rest?
> Snow smothered me,
> Rain smote me,
> Dew dropped on me,
> Long was I dead.

Odin demands to know Baldur's impending fate. Eventually the vala recognizes her disturber and the interview ends acrimoniously:

> *Wöla*
> Du bist nicht Wegtam
> Wie erst ich wähnte,
> Odin bist du,
> Der Allerschaffer.
> *Odin*
> Du bist keine Wöla,
> Kein wissendes Weib,
> Viel mehr bist du
> dreier

---

[56] e.g. by Cooke, *World*, p. 111; Golther, *Grundlagen*, pp. 79–80.
[57] Simrock, *Edda*, p. 38.

Thursen Mutter.[58]

The vala
You're not Vegtam
As I first thought,
You are Odin,
The universe-maker.

Odin
You're no vala,
No wise woman,
But rather mother
To three monsters.

In Wagner's scene Wotan, standing before the vala's vault-like cave, sings the wise-woman awake. He names himself only 'Der Weckrufer' ('Awakener') and asks for counsel. There is the same tussle of wills and the same rancorous recognition at the conclusion of the scene:

Wala
Du bist—nicht,
was du dich nennst!
Was kamst du störrischer Wilder,
zu stören der Wala Schlaf?
Friedloser,
lass mich frei!
Löse des Zaubers Zwang

Wanderer
Du bist—nicht,
was du dich wähn'st!
Urmütter Weisheit
geht zu Ende:
dein Wissen verweht
vor meinem Willen.[59]

The vala
You are—not
what you name yourself!
Wild one, why did you come
to disturb the wise woman's sleep?
Fractious one,
let me free!
Unbind the magic bond!

[58] Simrock, *Edda*, p. 39.
[59] *Schr.*, vi. 155–6.

Wanderer
You are—not
what you suppose!
Ancestress-wisdom
comes to an end:
your knowledge wilts
before my will.

Just as the scene's structure, purpose and, in part, dialogue originate in the *Vegtamskvidha*, so too does the figure of the vala, perhaps even more so than is generally realized. Not until *Das Rheingold* did Wagner devise for her the alternative and competing persona of Erda, the earth-goddess, which later usurped to some extent in the public eye the original image of the long-dead vala. As he first conceived her in the young Siegfried drama, Wagner's vala is quite the *Vegtamskvidha* grave-dweller: 'The Wanderer is standing in front of a vault-like cave entrance . . . The cave vault has started to grow light. In the bluish glow the Vala rises out of the ground; she looks as if she is covered in frost, hair and clothing exude a glittering lustre.'[60] Curiously, although in the revision from *Der junge Siegfried* to *Siegfried* Wagner changed the name to Erda and altered some of the dialogue, he did not amend his original description of the vala to bring it into line with the substantially different description in *Das Rheingold*.

Possibly Wagner was influenced in his account of the rising and, particularly, the sinking vala at the end of her interview with Wotan by a scenario Studach devised in the notes to his translation of the Edda, which Wagner had read back in Dresden. Studach was not actually writing of the vala in *Vegtamskvidha* but of her sister in the *Völuspá*, whom he likewise envisaged delivering her counsel to Odin from the grave: 'In the context of the poem the vala is speaking from the grave and not as a living woman . . .'[61] Possibly Studach regarded this as standard form for prophesying valas.

[60] 'Vor einem gruftähnlichen höhleneingange steht der Wanderer . . . die höhlengruft hat zu erdämmern begonnen: im bläulichen lichtscheine steigt die Wala aus der der erde, sie erscheint wie von reif bedeckt, haar und gewand werfen einen glitzernden schimmer von sich.' (*Skizzen*, p. 168.)

[61] 'Wola, im Sinn des Liedes, vom Grab aus redet und nicht als lebende Frau . . .' (*Edda*, p. 7 n.).

Certainly he reads the last line of the poem as referring to the vala sinking back into her grave, and notes: 'In order to be understood the whole poem needs to be read as though it came from the gates of the underworld and as though after completing her vision the prophetess then sank back again into the domain of death.'[62]

Whether or not this particular remark of Studach's influenced Wagner, other, equally imaginative, contributions from contemporary Nibelung scholars certainly did. Apart from the virulent new Edda strain in the mythological scenes of the Young Siegfried drama, there is a subtle but pervasive undercurrent of Romantic scholarship, sometimes elusive, sometimes enigmatic, which courses through the scenes and imbues them with ambiguities.

The decision to include a scene based on the *Vegtamskvidha* in his Siegfried drama may itself have been the result of Wagner's reading of the scholars. As the alternative title of the poem, *Baldrs Draumar*, suggests, this is a poem about Baldur, 'the white god, light god, shining like the sky, light, and day', as Jacob Grimm describes him.[63] The parallels between Baldur and Siegfried were obvious enough to scholars of Wagner's era: radiant young Odin-descendants, their invulnerability fatally flawed, both falling victim to an insidious murder plot, treacherously slain by the one-eyed Hagen or the blind Hödr.

Von der Hagen, writing in *Die Nibelungen*, unsurprisingly concludes from these parallels that Siegfried and Baldur are one and the same: 'Our Siegfried, made invulnerable by the dragon blood except for one concealed spot, is in fact quite the Scandinavian divine son Baldur.'[64] Lachmann was more cautious. In the final 'reduction' of the Siegfried saga in his 'Kritik' he arrives at a 'comparison (which, however, should

---

[62] 'das ganze Lied, wenn es verstanden werden soll, so gelesen werden will, als käm' es von den Thoren der Unterwelt und sie, die Seherin, sänke jetzt, nach Vollendung ihres Gesichtes, wieder hinunter ins Gefilde des Todes.'

[63] 'der weisse gott, lichtgott, der wie himmel, licht und tag leuchtende' (*Mythologie*, p. 203).

[64] 'Unser vom Drachenblut . . . bis auf eine verborgene Stelle unverwundbare Siegfried ist nämlich ganz der Nordische Göttersohn Baldur.' (p. 60.) Von der Hagen returns several times to the same theme, incorporating such material as the heightened funeral pomp for both victims and the similar properties of Siegfried's hoard and Baldur's ring Dröpnir (pp. 60–1, 68–9, 83–4, 148).

not be taken as a crude identification)'.[65] Siegfried, Lachmann argues, was originally the name of a god: 'If we accept this, then naturally he immediately brings to mind the Scandinavian god Baldur, as a god who likewise died.'[66]

Ashton Ellis writes of *Die Wibelungen* that for Wagner 'Friedrich was Siegfried, and Siegfried was Baldur, and Baldur was Christ.'[67] Obviously in such circumstances Wagner would have found it perfectly natural to include a Baldur scene in a Siegfried drama. Whatever the state in *Die Wibelungen*, however, by the time he reached *Der junge Siegfried* Wagner had moved on from any such 'crude identification' of Siegfried and Baldur. He was now thinking more in terms of a progressive association of the two. In the prose sketch for the vala scene of *Der junge Siegfried* the god Baldur is mentioned, in a sequence of ideas which link him with Siegfried but nevertheless retain his separate identity: 'The Wanderer: "The gods have been concerned about their end since Baldur sank, the loveliest of the gods; until then the world was at peace . . ."'[68]

In the poem for *Der junge Siegfried* Baldur's name was dropped; an allusion to the god however remained. Wagner achieved this by following a device he had originally found in Lachmann's 'Kritik'. Lachmann's strictures against 'crude identifications' did not extend to crude etymologies of the kind Wagner found irresistible; he himself made a play on the name Siegfried, 'a god of peace [Frieden] through victory [Sieg]',[69] which Wagner duly enshrined in the *Mythus* as 'Siegfried (who is to bring peace through victory)'.[70] Shortly afterwards he discovered in Weinhold's article on Loge in the first 1849 issue of the *Zeitschrift für deutsches Alterthum* the identical concept being used to link Siegfried with Baldur: 'Baldur is indeed a god of peace, but a Germanic peace god, a Sigufrit, who thrusts his way through to peace by victory,

---

[65] 'vergleichung (die aber keine rohe identification sein soll)' (pp. 344–5).

[66] 'Nehmen wir dies an, so denkt man bei ihm natürlich sogleich an den nordischen Baldur als einen gott der ebenfalls gestorben ist.' ('Kritik', p. 344.)

[67] Quoted in Newman, *Life*, ii. 21.

[68] 'd. Wand. "Um der götter ende sorgen die götter, seit Baldur sank der holdeste gott: so lange lag die welt in frieden . . ."' (*Skizzen*, p. 88).

[69] 'ein gott des friedens durch den sieg.' ('Kritik', p. 345).

[70] 'Siegfried (der durch Sieg Friede bringen soll)' (*Skizzen*, p. 28).

the sword.'[71] The 'peace through victory' tag could obviously unite the two Odin-offspring. Left as Weinhold had put it, however, the association was steering perilously close to 'crude identification'. To maintain the separate identity of Baldur and Siegfried while retaining Weinhold's unifying idea Wagner hit upon an ingeniously simple solution: he left the phrase as it stood for Siegfried, and for Baldur he inverted it. So in *Der junge Siegfried* Siegfried brings peace through victory, Baldur victory through peace.

> *Wanderer*
> Um der seligen ende
> sorgen die götter
> seit der erfreuende sank
> der im frieden siege schuf.
>
> .   .   .   .   .
>
> doch da nun kampfnoth
> kam in die welt,
> seit nur Sieg
> frieden noch schafft . . .[72]

> *Wanderer*
> The gods have been anxious
> about their end
> since the Bringer of Gladness fell,
> who brought victory through peace.
>
> .   .   .   .   .   .
>
> But now that strife
> has struck the world,
> since only victory
> still secures peace . . .

As a result of changes Wagner later made in the poem this whole passage was replaced and the 'Bringer of Gladness' vanished. In the present text of *Siegfried* there is no longer either reference or allusion to Baldur.

Another etymology, this time from Jacob Grimm's *Mythologie*, was the inspiration behind the atmospheric description of

---

[71] 'Baldur ist allerdings ein friedensgott, aber ein germanischer friedensgott, ein Sigufrit, der durch sieg, das schwert, zum frieden dringt.' ('Loki', p. 57.)

[72] *Skizzen*, p. 172. Strobel notes that 'der erfreuende' is Baldur by reference to the equivalent passage in the prose plan.

the Wanderer's approach at the opening of Act II. It is deepest
night, and Alberich is brooding alone outside Fafner's cave:

> In Wald und Nacht
> vor Neidhöhl' halt' ich Wacht:
> es lauscht mein Ohr,
> mühvoll lugt mein Aug'. —
> Banger Tag,
> bebst du schon auf?
> Dämmerst du dort
> durch das Dunkel her?
> *(Sturmwind erhebt sich rechts aus dem Walde.)*
> Welcher Glanz glitzert dort auf?
> Näher schimmert
> ein heller Schein;
> es rennt wie ein leuchtendes Ross,
> bricht durch den Wald
> brausend daher.
> Naht schon des Wurmes Würger?
> Ist's schon, der Fafner fällt?
> *(Der Sturmwind legt sich wieder; der Glanz*
> *verlischt.)*
> Das Licht erlischt—
> der Glanz barg sich dem Blick:
> Nacht ist's wieder.
> Wer naht dort schimmernd im Schatten?[73]

> In the wood by night
> at Neidhöhle I hold watch:
> my ears are pricked,
> my eyes strain to see. —
> Troubled day,
> are you trembling to life?
> Is that you dawning
> there through the darkness?
> *(A high wind rises right from the forest.)*
> What glow is that starting to glimmer?
> A shining brightness
> shimmers closer;
> it runs like a radiant steed,
> breaks through the wood,
> rushing this way.

[73] *Schr.*, vi. 123.

Is the worm's assailant approaching?
Is it Fafner's slayer already?
*(The wind drops again; the glow dies
away.)*
The light fades,
the lustre is veiled from view:
It's night again.
Who draws near, shimmering in shadow?

The 'worm's assailant' is still a long way off, wrangling along the route with Mime; Alberich's present visitor is not Siegfried but Wotan.

The analysis of names formed an important part of Grimm's method in the *Mythologie*. Since Odin/Wotan went under so many names the chapter on the god offered a lot of scope. Among the several Wotan names examined by Grimm is 'Omi' or 'Vôma', for which he gives the Latin definition, 'impetus, fragor [turbulence, uproar]'. Grimm develops the idea further: 'Omi, Vôma, however, could be conceived as an air-god . . . whose commotion through the heavens can be heard at daybreak, in the tumult of battle and in the cavalcade of the "wütendes Heer."'[74] Of the 'wütendes Heer', the ghostly army which rampages through the countryside at night, we shall see more in the following chapter. The Vôma aspect of Wotan is returned to further on by Grimm in his discussion of daybreak:

Even more significant and telling, however, are those expressions which associate with daybreak and dawn once more the idea of a vibration and a sound, which may be attributed to the wings of the approaching messenger of day but which in fact lead us to the highest godhead, whose activity sets the air in motion. Wuotan, conceived as Wuomo, Vôma, is a quivering of nature, manifested among other things at break of day when a fresh breeze blows through the clouds.[75]

[74] 'Man dürfte aber den *Omi, Vôma* sich als einen luftgott . . . denken, dessen rauschen am himmel bei tagesanbruch, im tossen der schlacht und im aufzug des wütenden heers vernommen wird.' (*Mythologie*, p.132.)
[75] 'Noch bedeutsamer und eingreifender sind aber die redensarten, welche mit tagesanbruch, mit morgenröthe wiederum die idee einer *erschütterung*, eines *geräusches* verbinden, das den schwingen des nahenden tagboten beigemessen werden darf, aber uns sogar zu dem höchsten gott führt, dessen walten die luft erschüttert. Wuotan als Wuomo, Vôma gedacht ist ein schauern der natur, wie es sich auch beim

The stiring of the air heralding the coming day was perhaps the most poetic of all the images of the god Grimm derived from his study of Wotan's names.

Like Grimm, Wagner was fascinated by the phenomenon of Wotan's many names, and already in the Young Siegfried drama he was finding new ones for him. In the Act I knowledge contest the Wanderer uses one such freshly coined name when he describes the gods to Mime:

> Auf wolkigen Höh'n
> wohnen die Götter;
> Walhall heisst ihr Saal.
> Lichtalben sind sie;
> Licht-Alberich,
> Wotan waltet der Schar.[76]
>
> On cloudy heights
> dwell the heavenly host,
> Walhall is their hall.
> Light-elves they,
> Light-Alberich,
> Wotan governs the gods.

At first sight 'Light-Alberich' appears to be a name with a respectable scholarly pedigree. Earlier on in the knowledge contest the Wanderer has spoken of the Nibelungs and their lord as 'Schwarzalben' and 'Schwarz-Alberich': 'black elves' and 'Black-Alberich'.[77] Wagner knew from the scholars' discussion of dwarf names that 'Alberich' meant 'lord of the elves' (see Chapter 5); he also knew, from Snorri, from Jacob Grimm, and from Ettmüller, that there were light-elves as well as black-elves.[78] To call the chief god 'Lord of the light-elves' by analogy with 'Lord of the black-elves' for the chief dwarf was therefore perfectly logical, always providing that the gods might be identified with the light-elves in the same way that the dwarfs were identified with the black-elves.

---

anbruch des tages erzeigt, wo frisches wehen durch die wolken dringt.' (*Mythologie*, p. 707.)

[76] *Schr.*, vi. 103–4.

[77] *Schr.*, vi. 102.

[78] Grimm, *Mythologie*, pp. 413–34; Grimm distinguishes a 3rd category, 'Dökkalfar' ('black-elves'), in addition to the other 2. See also Ettmüller, *Vaulu-Spá*, p. 47.

Dwarfs and black-elves posed no problems. Jacob Grimm had argued cogently in his *Mythologie* that the two were one and the same, concluding: 'The identity between black-elves and dwarfs must be upheld.'[79] No one, however, was prepared to do the same for the light-elves and the gods. Despite the frequent juxtaposing of 'alfar' ('elves') and 'aesir' ('gods') in the Eddas and other source material, scholars even of Wagner's day fought shy of suggesting that the two were one. Only partial assistance was available if Wagner was seeking reassurance for his initiative from some external authority: Ettmüller's reference to the gods in *Sigrdrífumál* as 'the race of light',[80] or Wilhelm Müller's broad definition of the term 'alfar':

If we wish to pursue the difference between these races, namely between the vanir and alfar and the aesir, along mythological lines, all that transpires is that 'alfar' has a more general meaning than 'aesir', since it can also be applied to spirits of a lesser order, such as the dwarfs for example.[81]

The elves, in other words, might be gods or dwarfs or indeed neither. This was the best help Wagner got.

It is probably better to see in Wagner's 'Light-Alberich' not a precise statement on the identity of gods and light-elves but rather a more general wish to use the light-versus-dark contrast to symbolize the major opposing power forces in his drama, along the lines described by Jacob Grimm: 'In the antithesis between light- and black-elves we find the dualism which other mythologies also establish between good and evil, friendly and hostile, heavenly and infernal spirits, between the angels of light and of darkness.'[82]

---

[79] 'Festgehalten werden muss die identität der *svartalfar* und *dvergar*.' (*Mythologie*, p. 415.)

[80] 'dem Geschlecht des Lichtes' (*Edda*, p. 21 n.).

[81] 'Wollen wir den unterschied dieser geschlechter, namentlich der Vanen und Alfen von den Asen auf dem mythologischen wege verfolgen, so ergibt sich nur, das *alfar* eine allgemeinere bedeutung hat als *aesir*, da dieser name auch geistern untergeordneten ranges, wie z.b. den zwergen zukommt.' (*Altdeutsche Religion*, pp. 178–9.)

[82] 'Man findet in dem gegensatz der lichten und schwarzen elbe den dualismus, der auch in andern mythologien zwischen guten und bösen, freundlichen und feindlichen, himlischen und höllischen geistern, zwischen engeln des lichts und der finsternis aufgestellt wird.' (*Mythologie*, p. 414.)

Wagner's interest in light-and-dark contrasts led him almost inexorably into the realms of solar myth. The Wanderer's Act II appearance as the prelude to daybreak can be seen as a sample. Certain of Wagner's other excursions into solar myth, however, have been marred by later changes to the text of the Young Siegfried drama, and have come down to us chiefly in the form of riddling passages which require a certain amount of restoration work before their former significance and interconnection come to light.

One such passage, the knowledge contest of Act I, occurs in the same scene as 'Light-Alberich' and relates likewise to Wotan. After the Wanderer has correctly answered the smith's questions and Mime's turn to be quizzed comes round, the dwarf muses doubtfully:

> Lang' schon mied ich
> mein Heimatland,
> lang' schon schied ich
> aus der Mutter Schoss;
> mir leuchtete Wotans Auge,
> zur Höhle lugt' es herein:
> vor ihm magert
> mein Mutterwitz.[83]

> I long since fled
> my fatherland,
> long since left
> my mother's lap;
> I was caught by Wotan's eye
> shining in at the cave:
> my mother wit
> waned before it.

Mime claims that his knowledge is shrivelling because of Wotan's eye. The most obvious, and perhaps most frequent, interpretation of Mime's brooding remark is that the appearance of the Wanderer at the entrance to Mime's cave has indeed robbed the frightened dwarf of his wits. But Wagner was thinking more in terms of that standard threat to dwarfs, the undoing of the knowledgeable Alvíss whom we encountered earlier in this chapter: exposure to sunlight.

[83] *Schr.*, vi. 105.

Since deserting his subterranean cavern for the open forest cave where he lives with his human charge, Mime finds his native wisdom wilting under the rays of the sun.

In the original text for *Der junge Siegfried* the subject of Mime's vulnerability was made clear in some lines the Wanderer delivers about himself—Wotan, 'Light-Alberich' —in his concluding answer to Mime's knowledge-contest questions:

> nur ein auge
> leuchtet an seinem haupt
> weil am himmel das andre
> als sonne den helden schon glänzt.[84]

> One eye only
> gleams in his head, because
> in the heavens the other
> shines as the sun on the heroes.

Der junge Siegfried had made it clear that the 'Wotan's eye' which so distresses Mime is the sun. As Wagner initially conceived his Young Siegfried drama there was no confusion; in his later recensions the Wanderer's lines on Wotan's eye were abandoned and Mime's observation, deprived of its referent, strikes the reader or listener as merely abstruse.

The notion of Wotan's eye as the sun was well beloved of Wagner's contemporaries. Already von der Hagen had referred to Odin as 'the one-eyed sun-god'[85] and written of the sky, 'in which Odin's one eye is present as the sun.'[86] Wilhelm Müller had similarly written in his chapter of Wotan: 'The very fact that people imagined him as one-eyed points to a sky-god, whose eye was thought to be the sun shining down on the earth.'[87]

It was unfortunate, therefore, as Jacob Grimm concedes, that the Eddic accounts of the creation of the universe do not bear these scholars out. Indeed, *Vafthrudhnismál* gives quite a different explanation of the sun's origin.[88] Grimm, however,

[84] *Skizzen*, p. 123.
[85] 'der einäugige Sonnen-Gott' (*Die Nibelungen*, pp. 69–70).
[86] 'an welchem Odins eines Auge als Sonne stehet.' (*Die Nibelungen*, p. 108.)
[87] 'Schon dass man ihn sich als einäugig vorstellte, weist auf den himmelsgott hin, als dessen auge die sonne gedacht wurde, welche die erde erleuchtet.' (*Altdeutsche Religion*, p. 183.)
[88] Ver. 23; see also *Völuspá*, ver. 5, and Snorri's account, Simrock, *Edda*, p. 246.

had unearthed a handful of sun–eye associations in early German texts,[89] and had marshalled evidence from so many other mythologies where the sun is an eye that the omission from the Eddas could easily appear an oversight: 'It is all the more noticeable that the Edda lacks the relevant relationship between the sun and the eye, since sun, moon, and stars are quite commonly supposed to be eyes.'[90]

Those who sought support from the Eddas for their solar myth interpretation of Wotan's lone eye took it from the dark *Völuspá* account of how the god came to lose his other one:

> Alles weiss ich, Odin,
> Wo du den Auge bargst:
> In der vielbekannten
> Quelle Mimirs.
> Meth trinkt Mimir
> Jeden Morgen
> Aus Walvaters Pfand.[91]

> I know all things, Odin:
> Where you hid your eye
> In Mimir's well,
> Of widespread fame.
> Mimir drinks mead
> Every morning
> From Walvater's pledge.

Wagner himself alludes to these *Völuspá* lines in the Norn Prologue of *Götterdämmerung*:

> Ein kühner Gott
> trat zum Trunk an den Quell;
> seiner Auges eines
> zahlt' er als ewigen Zoll . . .[92]

> A daring god
> went to drink from the well;
> one of his eyes
> he paid as eternal tribute . . .

---

[89] *Mythologie*, pp. 531–2.
[90] 'Desto mehr fällt auf, dass die Edda des treffenden verhältnisses der sonne zum auge entbehrt, da sonne, mond und sterne ganz gewöhnlich für augen gelten.' (*Mythologie*, p. 534.)
[91] Simrock, *Edda*, p. 6.
[92] *Schr.*, vi. 178.

On the *Völuspá* passage Simrock comments: 'This passage is usually interpreted as an allusion to Odin's lack of an eye; the sun is held to be Odin's one eye, the other its mirror image, however, reflected in the water at sunrise or sunset.'[93] Jacob Grimm was himself sufficiently persuaded of this theory to write in the *Mythologie* that the sun was

Wotan's eye to our ancestors. According to a story from the Edda Odin had to pledge one of his eyes to Mimir, or conceal it in his well, and this is why he is portrayed as one-eyed. With this eye the god surveys the whole world and nothing can remain hidden from his surveillance: he penetrates everything . . .[94]

All were agreed, then, that Wotan's eye was the sun. But which eye? The missing one, or the eye in Wotan's head? Wagner's mentors were overwhelmingly in favour of the latter explanation; Wagner himself chose the former:

> . . . am himmel das andre
> als sonne den helden schon glänzt.
>
> in the heavens the other
> shines as the sun on the heroes.[95]

It is Wotan's other eye that comes peering into the cave and unsettling Mime, not the one in his head. Wagner's only support for this came from Studach, who had written in the notes to his translation of *Völuspá*, at the mention of Odin's eye in the well, 'The vala . . . sees Odin's eye (the sun).'[96]

With the problem of Wotan's eye solved the second solar-myth enigma resolves itself. In Act III Siegfried and his divine ancestor finally come face to face. After a while Siegfried notices that under his low-brimmed hat the Wanderer lacks

---

[93] 'Gewöhnlich deutet man diese Stelle als eine Anspielung auf Odins Einäugigkeit und lässt die Sonne Odins Eines Auge sein, das andere aber deren bei Sonnenauf- oder Untergang im Wasser gespiegeltes Bild.' (*Edda*, p. 336.)

[94] 'unsern vorfahren auge Wuotans, und nach einer fabel der edda muste Odinn sein eines auge dem Mîmir zu pfand setzen oder in dessen brunnen bergen und darum wird er einäugig dargestellt. Mit diesem auge überschaut die gottheit die gesamte welt und nichts kann der spähenden verborgen bleiben, sie durchdringt alles . . .' (p. 665).

[95] *Skizzen*, p. 123.

[96] 'Wola . . . sieht Othin's Auge (die Sonne).' (*Edda*, p. 15 n.).

an eye, and he proceeds to comment on it in his own inimitable manner:

> Das schlug dir einer
> gewiss schon aus,
> dem du zu trotzig
> den Weg vertratst?
> Mach' dich jetzt fort!
> Sonst möchtest du leicht
> das and're auch noch verlieren.[97]

> Someone must have
> sent it flying
> for being too stubborn
> in stopping the way.
> Take yourself off now,
> or you might easily
> lose the other one also!

The Wanderer replies:

> Ich seh', mein Sohn,
> wo nichts du weisst,
> da weisst du dir leicht zu helfen.
> Mit dem Auge,
> das als andres mir fehlt,
> erblickst du selber das eine,
> das mir zum Sehen verblieb.[98]

> I see, my son,
> that despite knowing nothing
> you're quick to come up with an answer.
> With the eye of mine
> that I am missing
> you yourself catch sight of the one
> that's left for me to see with.

Wotan's missing eye has been passed on to Siegfried. Siegfried naturally thinks this is a huge joke,[99] but we can appreciate now why Wagner insisted, against the current grain of opinion, that the sun was Wotan's missing eye, not the one in his head. The sun-god in Wagner's cosmos is not

---

[97] *Schr.*, vi. 160.
[98] Ibid.
[99] 'Zum Lachen bist du mir lustig!' ('I regard you as good for a laugh!': *Schr.*, vi. 160.)

Wotan, though lord of the light-elves and harbinger of dawn, but Siegfried, that other Baldur, Wotan's heir. We recall that Alberich expected day and the dragon-slayer to arrive together in that atmospheric passage at the beginning of Act II. We can perhaps also feel a touch of sympathy for Mime, a sun-allergic dwarf, burdened with such a charge; small wonder that rearing Siegfried has left his tutor a master of the art of fear.

The sun-god concept of Siegfried was among Wagner's earliest and the one he in turn believed to be of the greatest antiquity. In *Die Wibelungen*, as Golther notes,[100] Wagner had written of 'the oldest significance of the myth, in which Siegfried is revealed as a light- or sun-god.'[101] The central cosmic event is the dragon fight, which shows 'the personified light- or sun-god conquering and slaying the monster of elemental, chaotic night.'[102] By killing the dragon Siegfried gains the hoard; but this incites the dragon's heirs, the powers of darkness, against him, as we have seen (Chapter 5): 'When light conquered darkness, when Siegfried slew the Nibelung dragon, he also won as fair booty the Nibelung hoard the dragon was guarding.'[103] With Siegfried's death the cycle comes full circle: 'Just as day finally succumbs to night once more, just as summer must finally yield to winter, so too is Siegfried finally slain in his turn.'[104]

Just as adherents of the cosmic-myth school had discovered in Wotan's missing eye an allusion to the sun, so too were they quick to spot in Siegfried the makings of a sungod. *Die Wibelungen* itself, with its main emphasis on the dragon-fight and the hoard, shows the mark of Lachmann's 'Kritik'. Lachmann too regarded the hoard as belonging to

---

[100] *Grundlagen*, pp. 94–5.

[101] 'der ältesten Bedeutung des Mythus, in welcher wir *Siegfried* als Licht- oder Sonnengott zu erkennen haben.' (*Schr.*, ii. 119.)

[102] 'den individualisierten Licht- oder Sonnengott, wie er das Ungetüm des chaotischen Urnacht besiegt und erlegt.' (*Schr.*, ii. 131.)

[103] 'Als das Licht die Finsternis besiegte, als Siegfried den Nibelungendrachen erschlug, gewann er als gute Beute auch den vom Drachen bewachten Nibelungenhort.' (*Schr.*, ii. 133.)

[104] 'Wie nun der Tag endlich doch der Nacht wieder erliegt, wie der Sommer endlich doch dem Winter weichen muss, ist aber Siegfried endlich auch wieder erlegt worden.' (*Schr.*, ii. 132.)

the powers of darkness; he likewise saw Siegfried's ownership of the hoard bringing him ultimately and fatally into
their power. For Lachmann the saga demonstrates 'how even
a glorious, radiant god, a god of peace through victory,
cannot with impunity kill the mysterious guardians of the
cold northern underworld and rob the dragon of the gold
belonging to the gods of darkness.'[105]

Lachmann concluded his research into the early form of the
saga with the words: 'Eleven years ago [1818] a predecessor
did indeed arrive by a far less exacting route at a god and,
what is more, at a sun-god Siegfried . . .'[106] The predecessor
thus ungently referred to was Mone, whose earlier Nibelung
work, the *Einleitung*, had appeared in 1818. Like von der
Hagen in his almost exactly contemporaneous *Die Nibelungen*
(1819), Mone anticipates Wagner's position in *Der junge Siegfried* by deriving Siegfried's sun-god role from his Odin
ancestry. Both scholars take as their starting-point the name
of the first of the Wälsung dynasty in the *Völsunga saga*,
Sigge, which reappears both in 'Siegfried' and as one of
Odin's names. Von der Hagen argues from this a generic
sun-god link throughout the Wälsung dynasty:

> Sigge, the ancestor of the Scandinavian Siegfried, is like Baldur a
> son of Odin who . . . is in a certain sense himself the one-eyed sun-
> god, father of the solstices . . . and who as the unconquered sun
> bears several names incorporating 'Siggi' or 'Sieg' . . . a name
> which also runs right through Siegfried's family as that of the
> sun-children.[107]

It is left to Mone this time to suggest that Odin and Siegfried
are actually one and the same: 'All the indications are that
Siegfried is very probably One with the Scandinavian Odin,
who was also in some respects the Scandinavian light-god.

[105] 'wie selbst ein herrlicher leuchtender gott, ein gott des friedens durch den sieg,
nicht ungestraft die geheimnissvollen wächter im kalten nordlichen todtenreiche
morden und das gold der nächtlichen götter dem drachen rauben darf.' ('Kritik',
p. 345.)

[106] 'Zwar ist schon vor elf jahren [1818] ein vorgänger auf weit bequemeren wege
zu einem gott und sogar zu einem sonnengott Siegfried gekommen . . .' (ibid.).

[107] 'Des Nordischen Siegfried Ahnherr Sigge ist, wie Baldur, ein Sohn Odins,
welcher . . . in gewisser Bedeutung selber der einäugige Sonnen-Gott ist, Vater der
Sonnenwenden . . . und mehre Namen mit Sigi oder Sieg führt, als unbesiegte
Sonne . . . welcher Name auch durch Siegfrieds ganzes Geschlecht, als der Son-
nenkinder, geht.' (*Die Nibelungen*, pp. 69–70.)

Siegfried shares the same name root as him in fact, for Odin too is called Sigge . . .'[108]

The 'Sigge' connection, which Mone owned came from his former collaborator and mentor Creuzer,[109] sparked off the whole of the mythological interpretation which assumes such importance in his *Einleitung*: 'His comment, that in this world Odin was called Sigge, was a ray of light to me and the origin of my whole mythological interpretation of the *Nibelungenlied*.'[110] Mone's treatment of the sun-god theory dominates the second half of his *Einleitung*. He takes as his departure point the four sun festivals: spring and autumn equinoxes, midsummer, and midwinter; and since his chosen source, the *Nibelungenlied* offers evidence of any description for only one of these, the exercise necessarily finds Mone at his most inventive.

The sun festival which finds mention in the *Nibelungenlied* is that of midsummer. Just before Siegfried is killed he and his queen are invited to a feast. The feast is at midsummer; therefore, Mone concludes, midsummer is the ancient festival of Siegfried's death. For the three remaining sun festivals which the *Nibelungenlied* fails to mention Mone has an answer too: just as Christianity has taken the original midsummer Siegfried festival and renamed it after John the Baptist, so too do other events in the Christian calendar mask Siegfried festivals. Behind the figures of the Archangel Michael at Michaelmas, the reborn son at Christmas, and the festival of the dragon-slayer St George in April lurks the old native sun-god.[111] In his concluding celebration of Siegfried, the 'sun-god' and 'god of day', Mone names him 'the eye of the world'.[112]

We cannot of course be positive that Wagner knew Mone's

---

[108] 'nach allen Anzeigen ist er [Siegfried] höchst wahrscheinlich Eins mit dem nordischen Othin, welcher auch in mancher Beziehung der skandinavischer Licht-gott war. Sigfrit hat nämlich einerley Namenswurzel mit ihm, denn auch Othin hiess Sigge . . .' (*Einleitung*, p. 76).

[109] *Einleitung*, p. v.

[110] 'so ward mir auch seine Aeusserung: dass Othin im Leben Sigge geheissen, ein Strahl des Lichtes und die Ursache meiner ganzen mythologischen Deutung des Nibelungen-Liedes.' (Ibid.)

[111] Mone, *Einleitung*, pp. 75–86; Mone complains that St George's Day, 23 April, has been moved a month out of line (p. 84).

[112] 'das Aug der Welt' (*Einleitung*, p. 86).

Einleitung, or even von der Hagen's *Die Nibelungen*. We do know, however, that Wagner read and valued Mone's later work, the *Untersuchungen*. Perhaps Mone's handling of solar myth in the *Untersuchungen* is another clue on the trail of the hitherto lost significance of this work for Wagner.

As a new element in the sun-god argument Mone now introduces the maiden. Whereas his *Einleitung* had been devoted specifically to the *Nibelungenlied*, in the *Untersuchungen* Mone's pursuit of solar myth leads him to *Das Lied vom Hürnen Seyfrid*, in which Siegfried releases the princess Chriemhilt from the power of her dragon captor. The dragon has snatched the maiden from her father's palace and now guards her on the Drachenstein. One Easter Day he turns into a man and informs his captive that five years thence he will become human permanently and marry her. Siegfried arrives in time to avert Chriemhilt's doom: only just in time, says Mone, and therefore the dragon-fight falls at the spring festival. The maiden Siegfried releases can be none other than the spring goddess, Easter, and the dragon, Winter: 'Siegfried (or whoever it was) [*sic*] does battle with the dragon of winter for Easter, who has been kept prisoner for six months; winter is vanquished and Siegfried marries Easter . . .'[113] By some ingenious arithmetic which it would take too long and be too complicated to explain, Mone has arrived at six months' captivity for the maid although the poem clearly states that she has been held for at least four years. Suffice it to say that Mone returns to the Easter connection of the dragon-fight later.[114]

It was at the maiden scene that Wagner returned to the sun-god theme in his Young Siegfried drama. *Die Wibelungen* apart, Wagner had not so far given much evidence of his adherence to solar myth. In *Siegfrieds Tod* any such suggestion had been limited to lighting effects: darkness gives way to radiant daylight whenever Siegfried appears; on his death night falls, while his attempt to disguise himself as Gunther takes place in lurid evening light.

---

[113] 'Sigfrit (oder wer es sonst war) kämpft mit dem Winter-Drachen um die 6 Monate lang eingesperrte Oster, der Winter wird besiegt, und Sigfrit vermählt sich mit der Oster . . .' (*Untersuchungen*, p. 170).
[114] *Untersuchungen*, p. 185.

Now, after the preliminary exchange with the Wanderer, Siegfried breaks through the flame barrier and arrives at the summit of Brünnhilde's rock. Darkness yields to limpid blue sky and bright daylight. Siegfried finds the sleeping maiden and kisses her awake; Brünnhilde opens her eyes and says:

Brünnhilde
Heil dir, Sonne!
Heil dir, Licht!
Heil dir, leuchtender Tag!
Lang' war mein Schlaf;
   ich bin erwacht:
   wer is der Held,
   der mich erweckt'?

Siegfried
(von ihrem Blicke und ihrer
Stimme feierlich ergriffen)
Durch das Feuer drang ich,
das den Fels umbrann;
ich erbach dir den festen Helm:
   Siegfried heiss' ich,
   der dich erweckt.

Brünnhilde
Heil euch, Götter!
Heil dir, Welt!
Heil dir, prangende Erde!
Zu End is nun mein Schlaf . . .[115]

Brünnhilde
Hail, O sun!
Hail, O light!
Hail, O shining day!
My sleep lasted long:
   I have awoken:
   Who is the hero
   who wakened me?

Siegfried
(solemnly affected by her gaze and her voice)
I broke through the fire
that burnt round the rock;
I breached your close-locked helmet:
   Siegfried it is
   who wakened you.

[115] *Schr.*, vi. 166–7.

Brünnhilde
Hail, O gods!
Hail, O world!
Hail, O radiant earth!
My sleep is now at an end . . .

It is often pointed out how much Wagner's lines owe to the
Eddic poem of the awakening valkyrie, the *Sigrdrífumál*:

[Sigurdrifa]
1. Was zerschnitt mir die Brünne?
Wie brach mir der Schlaf?
Wer befreite mich
Der falben Bande?

Sigurd
Sigmunds Sohn:
Eben zerschnitt
Das Wehrgewand
Dir Sigurds Waffe.

Sigurdrifa
2. Lange schlief ich,
Lange hielt mich der Schlummer,
Lange lasten Menschenloosse.
Odin waltete,
Dass ich nicht wuste
Die Schlummerrunen abzuschüteln.

3. Heil dir Tag,
Heil euch Tagessöhnen,
Heil dir Nacht und nährende Erde:
Mit unzorngen Augen
Schaut auf Uns
Und gebt den Sitzenden Sieg.

4. Heil euch Asen,
Heil euch Asinnen,
Heil dir, fruchtbares Feld![116]

[*Sigurdrifa*]
1. What slit my breastplate?
How broke my sleep?

---

[116] Simrock, *Edda*, pp. 168–9; the *Sigrdrífumál* connection is mentioned by e.g.
Golther, *Grundlagen*, pp. 71–2, and Cooke, *World*, pp. 74–6.

Who set me free
From the fallow bonds?

Sigurd

Sigmund's son:
Sigurd's weapon
Has just cut
Your coat of armour.

Sigurdrifa

2. Long did I sleep,
Long did slumber hold me,
Long does the human lot endure.
Odin disposed
That I should not be able
To shake off the runes of sleep.

3. Hail, O day!
Hail, O sons of day!
Hail, O night and nourishing earth!
Look on us
With anger-free eyes
And bestow on us sitting here victory.

4. Hail, O gods!
Hail, O goddesses!
Hail, O fruitful field!

Two peculiarities of Wagner's version, however, appear
strangely to have escaped notice. In the first place, the Edda
greeting is addressed equally to the powers of darkness and
light. If Wagner had not noticed this himself, it would have
been brought to his attention by Ettmüller, who noted in his
translation: 'Sigrdrifa includes in her entreaty for favour *all*
beings honoured as divine, whether they belong to the race
of light: the aesir—or to the race of darkness: the giants.'[117]
Wagner, on the other hand, omits all reference to night or

---

[117] 'Sigrdrifa umfasst also in ihrer Bitte um Gunst *alle* göttlich verehrten Wesen,
mögen sie nun dem Geschlechte des Lichtes, den Asen, oder dem Geschlechte der
Finsterniss, den Jöten, angehören.' (*Edda*, p. 21 n.) The equality between light and
dark in the valkyrie's address was even clearer in Wagner's other trans. of the Edda
than in Simrock's, since as Ettmüller points out, the 'nährende Erde' referred to in
ver. 3 is the daughter of night. Ettmüller's own translation of this line runs: 'Heil
dir, Nacht und Nährling!' ('Hail, O night and nursling!': p. 21), while the Grimm
brothers' version reads: 'Heil dir Nacht und Tochter (der Nacht)!' ('Hail, O night
and daughter (of night)!': p. 211).

darkness and limits Brünnhilde's greeting to the forces of light.[118]

Secondly, the stage action which Wagner originally conceived for this scene and which he retained as late as the *Sämtliche Schriften* runs as follows:

(Siegfried kisses her long and passionately. — He then leaps up, startled. — Brünnhilde has opened her eyes. — He looks at her in astonishment. They both remain for a while engrossed in their mutual gaze.)

Brünnhilde then breaks into her rapturous greeting:

> Heil dir, Sonne!
> Heil dir, Licht!
> Heil dir, leuchtender Tag![119]

> Hail, O sun!
> Hail, O light!
> Hail, O shining day!

— and so forth. In other words, Brünnhilde's greeting to the sun follows on immediately after she has been lost in contemplation of Siegfried, or indeed, we should perhaps conclude, occurs while she is still gazing on him, for in the continuation of the passage quoted above we saw Siegfried replying 'solemnly affected by her gaze'.

The scene this introduces is ablaze with light imagery. By a series of phrases reminiscent of Mone Siegfried becomes in turn 'hoard of the world' and 'life of the earth';[120] he is 'radiant offspring'[121] and 'joy of the world',[122] 'wakener of life, conquering light'.[123]

Act III of *Der junge Siegfried* sees solar myth in Wagner's opus at its zenith. Thereafter its hold on Wagner waned. Wagner never disavowed it or sought to eradicate it from his Sieg-

---

[118] It is all the more surprising that this point should have been missed by Golther, since he refers repeatedly in his study of this scene to the predominance of light imagery (*Grundlagen*, pp. 72, 74).

[119] '(*Er [Siegfried] küsst sie lange und inbrünstig. — Erschreckt fährt er dann in die Höhe. — Brünnhilde hat die Augen aufgeschlagen. — Staunend blickt er sie an. Beide verweilen eine Zeitlang in ihren gegenseitigen Anblick versunken.*)' (*Schr.*, vi. 166.)

[120] 'Hort der Welt'; 'Leben der Erde' (*Schr.*, vi. 172).

[121] 'leuchtender Spross' (*Schr.*, vi. 173).

[122] 'Lust der Welt' (*Schr.*, vi. 167).

[123] 'Wecker des Lebens, siegendes Licht' (ibid.).

fried drama, but in more than one instance we witness a retreat. The Wanderer's explanation of 'Wotan's eye' in the Act I knowledge contest was omitted to make room for fresh material, and the information was lost to the drama. In the performing version of *Siegfried* the stage directions accompanying the waking of the valkyrie have changed: '(. . . *Brünnhilde opens her eyes. Siegfried jumps up and remains standing in front of her. Brünnhilde slowly raises herself up to a sitting position. With solemn gestures of her raised arms she greets her returned perception of earth and sky.*)'[124] No longer does Brünnhilde gaze on her awakener as she greets the sun.

The decreasing conviction in Wagner's solar myth was probably due to a combination of factors. The psychological and philosophical shake-up which Wagner underwent in the years immediately after the completion of *Der junge Siegfried*, and which left him in the mood for *Tristan*, no doubt made solar myth appear at best unimportant, at worst positively undesirable. In practical terms, too, Wagner may have decided that too overt a presentation of Siegfried as sun-god was unworkable on stage.

A third cause which could well have joined with these in dampening Wagner's sun-god enthusiasm was the personal influence of Ettmüller. The Zurich scholar had stood back from the prevailing light-myth trend. In the introduction to his translation of *Vaulu-Spá* he distances himself from the Mone–Creuzer school.[125] Later in the same edition he subjects the published views of Studach, a dedicated light-myth adherent, to the kind of acidulous polemic the era favoured.[126]

Clearly, any growing reservations Wagner may have had about solar myth in his Siegfried drama would have received support and encouragement from Ettmüller. Wagner's drift away from solar myth is first visible in the reworking of his Young Siegfried drama late in 1852; our evidence for Wagner

---

[124] '(. . . *Brünnhilde schlägt die Augen auf. Siegfried fährt auf und bleibt vor ihr stehen. Brünnhilde richtet sich langsam zum Sitze auf. Sie begrüsst mit feierlichen Gebärden der erhobenen Arme ihre Rückkehr zur Wahrnehmung der Erde und des Himmels.*)' See e.g. Friends of Covent Garden edn., ed. W. Mann (London, 1964), 92–3.
[125] pp. xlvi–xlix.
[126] pp. 36–44.

consulting Ettmüller comes from the spring of that year (see Chapter 2). If Wagner did indeed discuss the subject of solar myth with him, it is evident in which direction his Zurich acquaintance would have advised him to go.

8

The *Ring*

THE two remaining works of the cycle, *Das Rheingold* and *Die Walküre*, proceeded more or less in harness. Already in October 1851 Wagner was writing of extending his Siegfried dramas to three, with an additional full-scale prologue.[1]

Accordingly, the beginning of November 1851 saw the first sketch for *Das Rheingold*, originally entitled *Der Raub des Rheingolds*. A sketch for the first two acts of *Die Walküre* followed on the middle of the month. Over the winter Wagner added to his sketches, notably to the character of Loge and the prehistory of the Wälsungs.

In March 1852 Wagner wrote the full prose plan for *Das Rheingold*. Then, during a period of about six weeks from the second half of May on, he produced the prose plan for *Die Walküre*, followed by the complete poem. In the middle of September Wagner took up *Das Rheingold* again and by early November had completed the poem. At the same time, from mid-October to mid-December, he was engaged on the alterations to *Der junge Siegfried* and, more especially, *Siegfrieds Tod*, to bring them into line with his new overall concept and make them ready for publication early in 1853.

It was in connection with *Die Walküre* that Wagner left a rare and directly contemporaneous bibliographical indicator. On 12 November 1851, poised to commit his first sketch of *Die Walküre* to paper, he wrote to Uhlig from Albisbrunn that letter we cited in Chapter 1, asking him to forward from the Royal Library at Dresden von der Hagen's translation of the *Völsunga saga*, which he needed 'for another quick look through'.[2]

Clearly Wagner had the *Völsunga saga* in mind in the writing of *Die Walküre*; a comparison with the *Mythus* of 1848

---

[1] Letter to Theodor Uhlig, 12 Oct. 1851, *Dokumente*, p. 57.
[2] 'zu einer abermaligen kurzen Durchsicht' (*Dokumente*, pp. 57–8).

reveals a great expansion of the *Völsunga saga* material in the drama. Equally clearly, it was Wagner's mental image of the *Völsunga saga*, his recollection of the saga, which was most influential, and not the text itself, for by the time the book arrived from Dresden the first sketch for *Die Walküre* was complete. Only the supplementary 'History of the Wäl-sungs'[3] which emerged during the winter stood to benefit from Wagner's second quick glance through the *Völsunga saga*.

As with the Young Siegfried drama, a good part of Wagner's inspiration for *Die Walküre* came from his wider reading around the saga subject. After Siegmund has described the perils of his childhood he tells Hunding and Sieglinde:

> uns schuf die herbe Not
> der Neidinge harte Schar.[4]
>
> The harsh band of the Neidings
> brought us in dire need.

The name of the Wälsungs' enemies was something Wagner had salvaged from his 'Wieland' dream, one of the many dramatic schemes which he had conceived during the writing of his Nibelung drama but which never progressed beyond the initial sketch.[5] Both the *Thidreks saga* and *Poetic Edda* name as Wieland's chief antagonist a King Nidung, or in Simrock's German rendering in the *Amelungenlied*, Neiding. He is a suitable hate figure, mean-spirited and treacherous.

Wagner would have found the name all the more satisfactory for the Wälsungs' enemies through its semantic connotations as expounded by Simrock and von der Hagen. Both had pointed to a significance in the name. In Simrock's *Amelungenlied*, when the two Dietrichs engage in combat Dietrich von Bern calls to his fleeing rival:

> So steige nun vom Rosse und miss die Kraft mit mir
> In offnem Streit, sonst sag ich, es ist kein Herz in dir,

[3] 'Geschichte der Walsungen', *Skizzen*, pp. 211–12.
[4] *Schr.*, vi. 8.
[5] 'Wieland der Schmied' (*Schr.*, iii. 178–206); see also *ML*, p. 444.

Und sollst ein Neiding heissen hinfort vor Jedermann.[6]

Now dismount from your horse and try your strength
    with me
In open combat, else I'll say your heart's not where it
    should be,
And you'll be called a Neiding henceforth by everyone.

For Simrock 'Neiding' is tantamount to 'coward'. Von der
Hagen, who regarded the name of the king in the Wieland
saga, Nidung, and the Norse word 'nidingr' as one and the
same and equivalent to the modern German 'Neidhart',
emphasizes rather the qualities of malice and cunning. In a
note to the *Thidreks saga* he writes: 'Neidhart, like the Norse
and Old German Niding, Nidung, is a general term for an
evil, cunning, envious person, an envier . . . and at the same
time a significant personal name (like the above-mentioned
Niding . . .)'.[7]

Hunding, another name with somewhat unappetizing connotations, comes from among the ranks of Siegmund's
slayers in the *Völsunga saga*. In Hunding Wagner has combined this role with that of Siegmund's other major adver-
sary in the *Völsunga saga*: his hostile brother-in-law, husband
of Sieglinde.

When the fugitive Siegmund arrives at Hunding's hearth it
is his twin sister Sieglinde who greets him. As yet the twins
have not identified each other, but Hunding, watching the
unknown stranger together with his wife, observes:

> Wie gleicht er dem Weibe!
> Der gleissende Wurm
> glänzt auch ihm aus dem Auge.[8]
>
> He's so like the woman!
> The glistening worm
> gleams out of his eye too.

The special, distinguishing quality of the Wälsung eyes is
something Lachmann remarks on in his 'Kritik', where he

---

[6] *Heldenbuch*, vi. 213.
[7] 'Neidhart ist, wie das Nordische und Altdeutsche Niding, Nidung, ein all-
gemeiner Ausdruck für böser, listiger, neidischer Mensch, Neider . . . und zugleich
bedeutsamer Eigenname (wie der obige Nidung . . .)' (*Thss*, i. 260).
[8] *Schr.*, vi. 6.

singles out Sigurd's 'gleaming Wälsung eyes'[9] and 'shining Wälsung eyes':[10] those eyes which draw the Wanderer's attention when faced with his grandson in Act III of *Siegfried*. Fafner in the *Völsunga saga* and *Poetic Edda* makes a comment on Siegfried's eyes which would seem to suggest that they are an inherited feature. Wagner would not have known the *Völsunga saga* passage, since von der Hagen omitted to translate it, but in Simrock's *Edda* he would have found:

> Klaräugiger Knabe,
> Kühn war dein Vater,
> Dem Ungebornen vererbt' er den Sinn.[11]
>
> Bright-eyed boy,
> Bold was your father;
> The trait passed on to his unborn son.

The sign of the worm or dragon[12] Hunding remarks in the twins' eyes is a sure indication of common Wälsung kinship. In the story of Sigurd Ormr í Auga the identity of Aslaug, daughter of Sigurd/Siegfried and Brünnhilde, is under doubt. As evidence she produces a son, also named Sigurd, in whose eyes the sign of the dragon proves his descent from Sigurd Fafner's slayer. The tale is told in *Ragnar Lódbrokssaga*, translated in volume V of von der Hagen's *Nordische Heldenromane*. Wagner, we rather concluded in Part I, did not know volume V; but von der Hagen supplies a resumé of the episode in the introduction to the *Völsunga saga*,[13] and the story is retold by Russwurm in his *Nordische Sagen*.[14]

In Wagner's poem the distinctive eyes which Siegfried inherits from his divine grandsire and which mark the twins' blood-relationship also unite brother and sister with their father Wotan, forming one of the major poetic images in the exploration and unravelling of their identity. At the wedding

[9] 'glänzende Völsungaugen' (p. 339).
[10] 'leuchtenden Walsungaugen' (p. 343).
[11] p. 162. Wagner also knew the passage from Grimm, *Edda*, p. 181 and Ettmüller, *Edda*, p. 14.
[12] The semantic range of 'worm' was very wide in the Germanic languages of the Dark and Middle Ages and conveniently covered dragons and snakes as well as our modern worms. Wagner observes the same tradition when he refers to Fafner as a 'Wurm', e.g. *Schr.*, vi. 123.
[13] pp. xiii–xv.
[14] pp. 177–95.

feast of her forced marriage to Hunding Sieglinde is consoled
by the entrance of a sword-bearing stranger:

> tief hing ihm der Hut,
> der deckt' ihm der Augen eines;
> doch des andren Strahl,
> Angst schuf er allen,
> traf die Männer
> sein mächt'ges Dräu'n . . .[15]

> Low hung his hat,
> it covered one of his eyes;
> but the beam of the other
> put fear into all
> when men encountered
> its powerful menace . . .

In the boldness of his gaze she recognizes her father:

> An dem kühnen Blick
> erkannt' ihn sein Kind . . .[16]

> By his daring glance
> his daughter knew him . . .

Now, as she probes his identity, Sieglinde discovers the same
feature in Siegmund:

> Deines Auges Glut
> erglänzte mir schon: —
> so blickte der Greis
> grüssend auf mich,
> als der Traurigen Trost er gab.[17]

> The glow of your eyes
> gleamed on me once:
> so gazed the old man
> on me in greeting,
> as solace he brought for my sorrow.

Siegmund for his part, challenged to reveal his father's name,
finds in Sieglinde the same radiant eye:

> dem so stolz
> strahlte das Auge,

---

[15] *Schr.*, vi. 14–15.
[16] *Schr.*, vi. 20.
[17] Ibid.

wie, Herrliche, hehr dir es strahlt,
der war Wälse genannt.[18]

        he whose eye
    so proudly shone
as sublimely yours shines in your splendour,
Wälse—was his name.

For Siegmund his father's name is Wälse. We know him as
Wotan. The whole of the first act of *Die Walküre* is a complex, mysterious web of hidden and concealed identities.
Much of the complexity was late maturing, belonging to the
ancillary sketches Wagner made over the winter and spring
of 1852 and taking final form only as he settled to the writing
of the full dramatic text.
Wotan's dual identity as the twins' father Wälse is one such
late offshoot. In Wagner's original *Mythus* the Wälsungs'
father is neither Wotan nor Wälse but some unnamed hero,
who we merely deduce to be of divine origin.[19] Wotan's only
specified role in the twins' conception is rather akin to that of
a consultant at a fertility clinic: he provides an apple and the
twins' mother conceives.[20] Wälse is not named at all.
By the time he wrote *Siegfrieds Tod* Wagner had worked
out a four-generation family tree for Siegfried. At this stage
Wotan was established as head of the family and Wälse,
father of Siegmund and Sieglinde, was his son. The *Völsunga
saga* gives the name of the twins' father as Völsung. That
Wälse, and not the eponymous form Völsung/Wälsung, was
the authentic name of the dynastic founder was something
Wagner owed to Lachmann: 'The saga has little of significance to relate about Sigmund's father Välse, as he is called in
the Anglo-Saxon poem *Beowulf* (in Scandinavia he is himself
already known by the patronymic, Völsungr).'[21]
Lachmann evidently found the Wälse of the *Völsunga saga* a
shadowy figure. When Wagner started to work on his Wälse
for *Die Walküre*, he in fact abandoned what little the *Völsunga*

[18] *Schr.*, vi. 21.
[19] *Skizzen*, p. 28.
[20] *Skizzen*, p. 27.
[21] 'Von Siegmunds vater Välse, wie ihn das angelsächsische Beovulfslied nennt
(im norden heisst er selbst schon patronymisch Völsúngr) weiss die sage wenig
bedeutendes zu erzählen.' (Lachmann, 'Kritik', p. 339).

saga tells of him and devised instead his own 'History of the Wälsungs' over the winter of 1851–2, in which Wälse's role is established much as it now stands in Siegmund's Act I recital.

Almost at the last minute Wagner made one of his most inspired moves, so full of pathetic potential for the drama: Wälse as Wotan's son disappeared, and the name became instead one of the many disguises under which the god himself moves among mortals. The first unambiguous indication of the new arrangement appears in the prose plan of May 1852 for Act II of *Die Walküre*, where Fricka reproaches her consort: 'But now, since you took to calling yourself Wälse and roaming the woods like a wolf . . .'[22] Wotan's new identity as the twins' father Wälse seems to have come to Wagner in conjunction with yet a third name for the god, and quite possibly as a result of it.

> Wolfe, der war mein Vater[23]

> Wolf, he was my father

—thus Siegmund begins his story when, trapped in his enemy's hall, he wishes to conceal his identity. From here on wolf imagery is rife in the history Siegmund tells. The twins' childhood home is 'the wolf's lair',[24] and when it is destroyed Siegmund and his father live in the forest as 'the wolf pair'.[25] The last sign Siegmund finds of his father is a wolfskin:

> eines Wolfes Fell
> nur traf ich im Forst:
> leer lag das vor mir,
> den Vater fand ich nicht.[26]

> A wolfskin was all
> I found in the wood;
> it lay there empty before me:
> my father I did not find.

Later, in Act II, Fricka will contemptuously refer to brother and sister as 'the she-wolf's litter'.[27]

---

[22] 'Doch jetzt, da dir Wälse zu heisen gefiel und wolfgleich in wäldern [du] chweiftest . . .' (*Skizzen*, p. 239).

[23] *Schr.*, vi. 7.

[24] 'das Wolfsnest' (*Schr.*, vi. 8).

[25] 'das Wolfspaar' (ibid.).

[26] *Schr.*, vi. 9.

[27] 'dem Wurfe der Wölfin' (*Schr.*, vi. 31).

Part of the inspiration for Wolfe and the prevailing wolf motif of Siegmund's boyhood narrative comes from within the *Völsunga saga* itself, where Siegmund and his son Sinfjötli, in hiding from their enemies, don wolfskins and live as marauding werewolves in the forest. Perhaps the stark mood of this chapter of the *Völsunga saga* was among the impressions Wagner wished to recapture when he asked for the saga to be sent to him in the autumn of 1852.

Again, though, Wagner called on inspiration from further afield. At a certain point in his recital Siegmund announces to Hunding:

> Ein Wölfing kündet dir das,
> den als Wölfing mancher wohl kennt.[28]
>
> A Wölfing is telling you this,
> whom as Wölfing many know well.

The Wölfings of German heroic literature traditionally belong to the circle of Dietrich von Bern, Siegfried's main rival for the title of leading hero, whose adventures were based on the North Italian homeland of certain Germanic tribes. The fantastical nature of Siegmund's disguised history has enabled Wagner to digress into the Dietrich saga, and in fact to the very beginning of the cycle. The wolf interest in *Die Walküre* has strayed from the all-male werewolves of the *Völsunga saga* over on to the maternal side: Siegmund and Sieglinde are 'the she-wolf's litter', their home the 'wolf's lair'. Evidently Wagner was thinking of the poem *Hugdietrich*, which tells how Wolfdietrich came to be born and how the Wölfings got their name.

However tragic Siegmund's childhood might be, Wagner's source for the twins' infancy was essentially comic: the poem *Hugdietrich* forms a light-hearted prelude to the more sombre *Wolfdietrich*. In the former poem, Hugdietrich disguises himself as the Princess Hildegund and goes to visit Princess Hildburg in her well-guarded tower. He becomes her playmate, and by the time he leaves Hildburg is pregnant. The distraught princess, fearing discovery, lowers her new-born son outside the castle walls, where a passing wolf takes him home to her lair. All ends well: the child is found playing among the cubs of his surrogate mother and is given the name Wolfdietrich. His finder, Berchtold, takes three wolves in his shield, and that is how his descendants come to be called 'Wölfings'. Wolfdietrich is restored to his parents, Hugdietrich marries his Hildburg, and they live happily ever after, for a time.

Numerous factors may have been in Wagner's mind when he decided to include in his Wälsung history material from the saga of the Wölfings. For a while in the *Poetic Edda* cycle, during the Helgi poems, the Wälsungs are themselves referred to as 'Ylfingar', which the Grimm brothers translate in their prose retelling (but not in their verse translation) as 'Wölfinge' throughout. 'Siegmund's descendants are called the Wälsungs or Wölfings', they comment on *Helgakvidha Hundingsbana fyrri.*[29]

How the Wälsung-Wölfings of the Helgi poems relate to those of the Dietrich cycle is another matter, and one over which there was considerable divergence of opinion among Wagner's contemporaries. Mone in his *Untersuchungen* painstakingly constructs a bridge between the Wälsungs of the north and the Wölfings south of the Alps, via the Bavarian Ilsungs: 'Ilsung was no other than Wilsung, and was the same not only as Welsung, Völsung but also as Wilz. The Bavarians must have known that Wils or Ils means 'a wolf', because they translated it as "Wölfing".'[30] Göttling could see no connection. In his *Nibelungen und Gibelinen* he singles out the poems such as *Rabenschlacht* and *Rosengarten*, where Dietrich and Siegfried meet in combat, as evidence of irreconcilable opposition between their respective factions.[31]

Von der Hagen, on the other hand, saw the strongest argument for the unity of the two branches of Wölfings precisely in Siegfried and Dietrich, 'whose unity with Siegfried is demonstrated in many ways . . .'[32] As Schneider

---

[29] 'Die von Siegmund abstammten heissen die Wolsungen oder Wölfinge' (*Edda*, pt. 2, p. 24).

[30] 'Ilsung war nichts anders als Wilsung, und dies sowohl mit Welsung, Völsung als auch mit Wilz einerlei. Die Baiern müssen noch gewusst haben, dass Wils oder Ils ein Wolf heisst, weil sie es mit Wölfing übersetzten.' (p. 21).

[31] *Nibelungen und Gibelinen*, pp. 88–95.

[32] 'dessen Einheit mit Siegfried sich vielfältig zeigte . . .' (*Die Nibelungen*, p. 138).

points out,[33] von der Hagen elsewhere in his *Die Nibelungen* calls Siegfried's grandfather 'Wolfung',[34] and while it is clear from the context that this is a misprint for 'Wolsung', it may none-the-less have influenced Wagner. Of more immediate interest is the parallel von der Hagen presents between the wolf histories of Siegfried's family and that of Dietrich, descendant of the child Wolfdietrich:

Dietrich's grandfather, called Wolfdietrich because a she-wolf suckled him when he was abandoned ... gives Hildebrand three wolves for his coat of arms: his family are known accordingly as the Wölfings ... And the Wälsungs are also known as Wölfings (Ylfinger); their dynastic founder, Sigi, is called a wolf (Warg) on account of a murder, and Siegmund, Siegfried's father, lives for a while as a werewolf.[35]

More than any of these, however, it is to Simrock that we owe the wolf imagery of *Die Walküre* and Wotan's triple identity. From our examination of the *Heldenbuch* literature in Wagner's library (Chapter 2) it emerges that while this did include *Wolfdietrich*, in the original-language edition produced by von der Hagen and Primisser, it did not include the poem which tells of Wolfdietrich's infancy, *Hugdietrich*. Instead Wagner found the slightly idiosyncratic version that Dietrich's wife Herrat tells to entertain the company on their homeward journey at the close of Simrock's *Amelungenlied*.

Simrock's concern was to show that Dietrich and the Amelungs were of divine descent. Knowing Odin/Wotan to be equally capable of Hugdietrich's style of courtship,[36] Simrock arranges a preliminary meeting between god and mortal in which the two change places. It is therefore Odin who, alias Hugdietrich, alias the fair Hildegund, visits Hild-

---

[33] 'Altertum', p. 113.

[34] p. 62.

[35] 'Dietrich's Grossvater, Wolf-Dietrich genannt, weil er, ausgesetzt, von einer Wölfinn gesäugt worden ... gibt Hildebranden drei Wölfe zum Wappen; und darnach heisst sein Geschlecht die Wölfingen ... Und Wölfingen (Ylfinger) heissen auch die Wolsungen, deren Stammvater Sigi, wegen eines Mordes ein Wolf (Warg) heisst, und von denen Siegmund, Siegfrieds Vater, eine Zeit als Wehrwolf lebt.' (*Die Nibelungen*, p. 102).

[36] *Heldenbuch*, vi. 417; Simrock was acquainted with Saxo's account of Odin's courtship of Rinda.

burg in her tower and begets Wolfdietrich, while Hugdietrich amuses himself in Walhall.

Simrock undertook his saga deviation with the parallel
Wälsung genealogy specifically in mind. In the appendix to
his *Heldenbuch* he explains that Odin's assumed identity was
introduced 'because I wanted to derive the Amelung family
from Odin, just as happens in the *Völsunga saga* with that of
the Wälsungs, from which Siegfried is descended . . .'[37] The
cross-influence from the Wälsung saga to the Dietrich saga
could work in the opposite direction too, and out of
Simrock's Aristophanic comedy of disguise was born
Wagner's poignant mystery of Wotan–Wälse–Wolf and the
poetic imagery of the Wälsung childhood.

Among Wagner's further source material Wolfgang Golther
cites 'two marvellous Scandinavian skaldic poems' as con-
tributing to the Death Annunciation scene in Act II of *Die
Walküre*.[38] Which two skaldic poems Golther does not specify, but from his treatment of the scene we can deduce that
one of the poems he intended is the *Hákonarmál*.[39] In the thick
of the fray the valkyries Göndul and Skögul appear on the
battlefield. King Hakon hears them talking:

> Göndul sprach
> Im goldnen Helm:
> 'Grösser wird die Zahl der Götter;
> Den starken Hakon
> Und all sein Heer
> Haben sie zu sich geladen.'[40]

> Göndul spoke
> In her golden helm:
> 'Greater will be the gods' assembly;
> The mighty Hakon
> And all his host
> Have been invited to join them.'

Hakon is not pleased. Being 'invited to join' the gods means

---

[37] 'weil ich das Geschlecht der Amelungen von Odin herleiten wollte, wie diess
auch in der Völsungasaga mit dem der Welsungen geschieht, aus welchem Siegfried
hervorging . . .' (vi. 417).
[38] 'zwei wundervolle nordische Skaldenlieder' (*Grundlagen*, p. 13).
[39] *Grundlagen*, pp. 56–7.
[40] Mohnike, *Heimskringla*, i. 148.

effectively the death sentence for him and all his men. He
protests to Göndul:

> 'War ich, Walkyria,
> Werth nicht des Sieges:
> Warum entscheidest du so die Schlacht?'[41]

> 'Was I, valkyrie,
> Not worthy of victory?
> Why did you so decide the battle?'

The valkyrie explains that though Hakon and his men fall in
battle, victory is nevertheless theirs:

> 'König, wir sandten
> Sieg dir von oben;
> Haben in Flucht geschlagen den Feind.'[42]

> 'King, from on high
> We sent you conquest:
> We have put your foes to flight.'

Skögul now chimes in:

> 'Reiten jetzt lasst uns
> Hin zu der Götter grünender Au;
> Dass wir Bölwerkern
> Bringen die Kunde,
> Hakon erscheine, um Odin zu schaun.'[43]

> 'Let us now ride
> Along to the green-springing field of the gods,
> So we may bring
> The news to Bölverk [Odin],
> That Hakon will enter to look at Odin.'

The poem concludes with Hakon's reception in Walhall.

Golther's other skaldic poem might be either the *Eiríksmál*,
where King Eirik arrives with great noise in Walhall, or the
song the dying Ragnar Lodbrok sings in prison:

> Walkyrien winken,
> Die Odin mir sendet
> Vom Saale der Götter.
> Auf dem Thron mit den Asen

---

[41] Mohnike, *Heimskringla*, i. 148.
[42] Ibid., 149.
[43] Ibid.

Soll freudig ich trinken.[44]

Valkyries beckon,
Sent me by Odin
From the hall of the gods.
On the throne with the aesir
I'll drink, full of joy.

The *Hákonarmál* was included in Wagner's preferred
translation of the *Heimskringla*, that by Mohnike.[45] A second
translation was available in the book by Frauer,[46] who also
gives the *Eiríksmál*, both in the original language[47] and in
German,[48] and the Ragnar Lodbrok poem.[49] Wagner knew
the latter also in Russwurm's translation,[50] and that of Göttling in his book *Ueber das Geschichtliche im Nibelungenlied*,
which Wagner borrowed from the Royal Library early in
1849.[51]

All this dates back to Wagner's Dresden days. We cannot
therefore agree when Golther suggests that it was Ettmüller
in Zurich who introduced Wagner to the skaldic poems.[52]
Nevertheless, the period of *Die Walküre* is one for which we
have some evidence of composer and philologist conversing
on such issues (see Chapter 4), and even if we cannot credit
Ettmüller with introducing Wagner to the poems he may
have been responsible for Wagner's decision to exploit them
in the Death Annunciation scene. Such an interpretation of
events is all the more likely when we consider Ettmüller's
own particular emphasis on the valkyrie role. Of all the
valkyrie functions mentioned in the literature—riding to battle, choosing the slain, serving in Walhall, and so forth—
none seemed to Ettmüller more uniquely a function of the
valkyries than the summoning of doomed warriors to
Walhall. In his *Vaulu-Spá* he writes: 'Valkyries are always

[44] Russwurm, *Nordische Sagen*, p. 221.
[45] pp. 146–50.
[46] *Die Walkyrien*, pp. 8–11.
[47] pp. 87–8.
[48] pp. 5–7.
[49] p. 4.
[50] *Nordische Sagen*, p. 221. All the above were in Wagner's personal library at
Dresden: see ch. 2.
[51] p. 71.
[52] *Grundlagen*, p. 13.

manifestly valkyries when they come to summon to Walhall the hereos whose death Odin has pre-ordained . . .'[53]

Golther's 'marvellous skaldic poems' apart, a multitude of references to the valkyries would have already been known to Wagner from the Edda. For a coherent presentation of the Eddic material, further development of its ideas, and additions from other sources Wagner turned to Ludwig Frauer's monograph *Die Walkyrien* and, as so often, to Jacob Grimm's *Mythologie*.

These were Wagner's main sources for most of the rest of his portrait of valkyrie life. Just occasionally, however, he includes a feature that owes its existence to none of these. One such occasion from the Death Annunciation scene is the passage where Brünnhilde tries to convince Siegmund that because he has seen her he must die:

> *Brünnhilde*
> Nur Todgeweihten
> taugt mein Anblick:
> wer mich erschaut,
> der scheidet vom Lebens-Licht.[54]

> *Brünnhilde*
> Only the fey
> behold my face:
> the person who sees me
> departs from the light of life.

There is no suggestion in the Eddas that gazing on a valkyrie is fatal; indeed, some heroes contrive to do just that for a good many years with no apparent ill effects. Perhaps Wagner was taking the position of Ettmüller and the skaldic poets on valkyrie appearances to doomed warriors a stage or two further; perhaps he was remembering Guttorm's description of the valkyries in Fouqué's *Sigurd* and drawing his own conclusions:

> Sie zieh'n
> Den Wahlplatz erst hindurch, zu küren sich,

[53] 'Walküren erscheinen überall als Walküren, wo sie die Helden, denen Othin den Tod schon bestimmt, nach Walhall zu laden kommen . . .' (p. 52).

[54] *Schr.*, vi. 49.

Wer im ruhmvollen Streite fallen soll.
Und wen sie küren, der erblickt alsbald
Ihr leuchtend Antlitz; freud'gen Schreckens voll
Bricht er durch Todesnacht in Walhall's Säle.[55]

First

They range across the battleground, to chose
Who it shall be that falls in glorious fight.
And straight away the chosen man perceives
Their shining countenance; with trembling joy
He breaks through death's dark night to Walhall's mansions.

Ettmüller was probably the ultimate source of another valkyrie duty which, Golther points out, has no authority in the Eddas: accompanying the fallen warriors to Walhall.[56] The immediate source, however, as Golther suggests, is just as likely to have been Jacob Grimm: 'The valkyries, however, ride to war, bearing with them the outcome of the battle and accompanying the fallen men to heaven . . .'[57] The idea of a valkyrie escort service enjoyed a certain currency at the time. Both Wilhelm Müller in his *Altdeutsche Religion*[58] and Ludwig Frauer in *Die Walkyrien*[59] include 'geleiten' (escorting) in their list of valkyrie functions. Probably both were drawing on Grimm. Prior to Grimm, however, Ettmüller had already numbered among the valkyries' tasks in his *Vaulu-Spá* 'accompanying the heroes to Walhall'.[60]

Perhaps, then, the corpses that bob from the valkyries' mounts were Ettmüller's contribution to the grand spectacular with which Wagner opens Act III of *Die Walküre*. Grimm helped with his mention of the valkyries' armour,[61] their riding through air,[62] and their association with clouds.[63]

Ever since 1848 Wagner had been eager to include an airborne entry for the valkyries in his drama. Gustav Freytag records a conversation with Wagner from September of that

---

[55] Fouqué, *Sigurd*, p. 187.
[56] *Grundlagen*, p. 57.
[57] 'Die valkyrien aber reiten in den krieg, bringen des kampfes entscheidung und geleiten die gefallenen gen himmel . . .' (*Mythologie*, p. 393).
[58] p. 199.
[59] p. 4.
[60] 'die Helden nach Walhall zu geleiten' (p. 56).
[61] *Mythologie*, p. 389.
[62] *Mythologie*, pp. 305, 398, 607.
[63] *Mythologie*, p. 306.

year in which the composer spoke of valkyries and of the ambitious stage effects he had in mind.[64] The first result was the valkyrie scene in Act I of *Siegfrieds Tod*, where the massed valkyries hover around Brünnhilde's rock and converse with their fallen sister in chorus.[65] The scene was later fined down to the present Waltraute scene, while the grand entry was reserved for *Die Walküre*, Act III. As Gerhilde, Ortlinde, Waltraute, and Schwertleite attend, clad in full armour, Helmwige, Siegrune, Grimgerde, and Rossweisse enter, one by one and two by two, after this fashion: 'A flash of lightning erupts in a passing cloud; in it a mounted valkyrie becomes visible; over her saddle hangs a slain warrior.'[66] The cloud disappears off right into the pine-trees, where the newly arrived valkyries dismount.

Apart from those of Siegrune and, of course, Brünnhilde herself, who are found in the Eddas, the valkyries' names are Wagner's own invention. This was despite the fact that Frauer had supplied over two dozen valkyrie names in a section devoted especially to their study.[67] Probably by 1852 Wagner could no longer remember them. He did, however, remember the principle that Frauer had said was at work in the valkyrie names: 'that war and martial life ... were regarded as personified in the valkyries',[68] and his own names are all composed of elements relating to the valkyries' military calling.

After the last cloud had delivered its load Helmwige observes:

> Acht sind wir erst:
> eine noch fehlt.[69]
>
> Eight are we only:
> One is still missing.

---

[64] *Erinnerungen*, quoted in Ellis, *Life*, ii. 261.

[65] *Schr.*, ii. 183–6.

[66] 'In einem vorbeiziehenden Gewölk bricht Blitzesglanz aus: eine Walküre zu Ross wird in ihm sichtbar: über ihrem Sattel hängt ein erschlagener Krieger.' (*Schr.*, vi. 58.)

[67] *Die Walkyrien*, pp. 33–6.

[68] 'das Krieg und kriegerisches Leben ... in den Walkyrien personifiziert angeschaut wurden' (*Die Walkyrien*, p. 35).

[69] *Schr.*, vi. 61.

Nine may have been the total number of Wagner's valkyries, but it was not a settled matter in his sources. He presumably preferred nine because a chorus of eight plus the soloist (Brünnhilde) appealed to him on grounds of stage effectiveness and vocal ensemble. Nine valkyries had some backing from Grimm, who writes: 'Usually nine valkyries ride out together . . .'.[70] But Grimm himself mentions other numbers, and elsewhere there was no unanimity.

Nor was there one single answer to the question of the valkyries' mythological position. Sometimes they appear as semi-divine beings, akin to norns, of supernatural origin. At other times they are daughters and sisters of kings, with family trees and human histories. Both Grimm and Frauer saw in the existence of two types of valkyrie an evolutionary development, a progression from immortal to mortal among the valkyries, whereby 'the oldest and most famous were descended, like the norns, from gods and elves'.[71] The more recent kind are likewise

full valkyries, differing from the earlier ones only in that they attach themselves to specific heroes in more human, womanly fashion, stand by them through thick and thin and intervene in their lives with all the energy and versatility which on the one hand their superhuman nature makes possible and to which, on the other hand, their womanly sentiment, the most tender, fervent, and noble love, drives them.[72]

According to what Act II of *Die Walküre* tells us, Wagner's valkyries are of the more ancient and austere order, divinely descended from Heervater—Father of Armies—himself. Wotan describes to Brünnhilde how she sprang from the union of highest god and earth mother:

in den Schoss der Welt

---

[70] 'Gewöhnlich reiten neun valkyrur zusammen aus . . .' (*Mythologie*, p. 392).

[71] 'die ältesten und berühmtesten, gleich den nornen, von göttern und elben stammten' (Grimm, *Mythologie*, p. 396).

[72] 'vollkommene Walkyrien, nur darin sich von den früheren unterscheidend, dass sie menschlich-weiblicher sich an bestimmte Helden anschliessen, ihnen in Noth und Tod beistehen, und mit all der Lebendigkeit und Vielseitigkeit in ihr Leben eingreifen, die ihnen einerseits ihre übermenschliche Natur möglich macht, und zu der sie andererseits ihr weiblicher Sinn, die zarteste, innigste und grossartigste Liebe treibt.' (*Die Walkyrien*, p. 53.)

> schwang ich mich hinab,
> mit Liebes–Zauber
> zwang ich die Wala,
> stört' ihres Wissens Stolz,
> dass sie nun Rede mir stand.
> Kunde empfing ich von ihr;
> von mir doch barg sie ein Pfand:
> der Welt weisestes Weib
> gebar mir, Brünnhilde, dich.[73]

> I let myself down
> into earth's lap;
> with love's magic
> I mastered the vala,
> disturbed her proud understanding,
> made her respond to me.
> Knowledge I took from her;
> but a token she got from me:
> the world's wisest woman
> bore for me, Brünnhilde, you.

Brünnhilde's companions in arms are equally high–born. We cannot deduce much from the evidence of Wotan, who later refers to Brünnhilde's 'eight sisters',[74] since according to Grimm 'these higher beings are always called "sisters" ',[75] but Fricka in her interview with her consort refers to the valkyries accusingly as

> . . . den schlimmen Mädchen,
> die wilder Minne
> Bund dir gebar . . .[76]

> . . . those ill–bred girls,
> the offspring of
> your illicit love . . .

By origin, therefore, Brünnhilde and her sisters belong at the beginning of the evolutionary process defined by Frauer and Grimm. Yet who can read Frauer's description of the younger branch, the 'more human, womanly' valkyries of

---

[73] *Schr.*, vi. 38.
[74] 'acht Schwestern' (ibid.).
[75] 'diese höheren wesen überall schwestern heissen' (*Mythologie*, p. 395).
[76] *Schr.*, vi. 30.

more recent development, without recognizing in it Brünnhilde as she appears from the end of Act II of *Die Walküre*
onwards?

The progression from divine to human was one which
interested Wagner deeply. He had already attempted the
theme once before: in the thick of the civil turmoil of 1849 he
had sketched out an Achilles drama, the main theme of which
was the hero's renunciation of the offer of immortality in
favour of the responsibilities of human friendship. Curt von
Westernhagen, writing of this period in his biography of
Wagner, remarks how the Achilles sketch foreshadows
Brünnhilde's action in the *Ring*.[77]

With Grimm and Frauer the development was evolutionary; in Wagner evolution has become revolution. One compassionate impulse in the Death Annunciation scene severs
Brünnhilde's divine ties for a mortal destiny and ushers in the
change-over from ancient to modern order of valkyries. The
'unfeeling maid'[78] has become 'the suffering, self-sacrificing
woman'.[79] So world-redemptive did Wagner consider this
process that he named a whole evening of his tetralogy in
honour of the valkyrie who underwent it.

Brünnhilde's metamorphosis takes place within the context
of what was generally agreed, Ettmüller apart (see above), to
be the primordial valkyrie role, implicit in the term itself:
choosing the slain. Both the 'val' and the 'kyrie' elements lent
themselves to this interpretation, though the precise significance of 'val', an element which recurs in 'Walhall' and in
Wotan's name 'Walvater', was open to debate. In his
Mythologie Jacob Grimm gives as the meaning of 'val',
'deposition of the corpses on the battlefield'; in a note to the
same page he further derives 'valr' or 'wal' from 'velja,
valjan—wählen [choose]'.[80] In the prose section of their *Edda*
the Grimm brothers refer to the species as 'Wahlküren—'

---

[77] *Wagner*, p. 139. Westernhagen actually suggests that *Siegfried* Act III is the
moment of Brünnhilde's transformation. The sketch is in *Schr.*, vol. xii.

[78] 'fühllose Maid' (*Schr.*, vi. 53).

[79] 'das leidende, sich opfernde Weib' (letter to August Röckel, 25 Jan. 1854,
Dokumente, p. 92).

[80] 'wal—niederlage der leichen auf dem schlachtfeld'; 'velja, valjan—wählen' (p.
389).

'choice-choosers'—throughout, emphasizing the 'choosing' aspect rather than the 'slain'. But whether the 'val' element signifies corpses on the battlefield, or itself indicates choice, or whether indeed, as Grimm suggests, the one concept derives from the other: the 'kyrie' element admitted of no other interpretation but 'choosing'. The second part of the name 'valkyrie', Jacob Grimm informs his readers, is associated with 'kiosa—kiesen [choose]'.[81] Ludwig Frauer echoes Grimm's analysis and concludes: 'The literal meaning of "valkyries" is therefore "selectors of those who are to fall".'[82]

The valkyries are therefore 'choosers'. But what sort of choice do they exercise, and above all, whose choice? Selecting which of the fallen heroes should go to Walhall evidently lies within the valkyries' choice, though it was generally agreed that it did not stop there. Frauer's definition—'selectors of those who are to fall'—indicates that he regards them as ruling over life and death in battle. Grimm suggests that their power extends over victory and defeat: 'deciding the issue of battle and victory is left in their hands.'[83] This brings us back to Göndul, Skögul, and King Hakon of the *Hákonarmál*; it is also a role we find Wagner attributing to his valkyries:

> Brünnhilde stürme zum Kampf,
> Dem Wälsung kiese sie Sieg![84]

> Let Brünnhilde advance to the fray;
> For the Wälsung she shall choose victory!

To the question of whose choice the sources give no clearcut answer. Sometimes the choice is evidently Odin/Wotan's: how else could Brünnhilde be punished for disobedience in the *Völsunga saga* or the Eddic *Helreidh*? At other times, as in the *Hákonarmál*, the valkyries appear to be acting independently. To conclude from the above evidence,

[81] Ibid.
[82] 'Valkyriur bedeutet also wörtlich: Auswählerinnen derer, die fallen sollen.' (*Die Walkyrien*, p. 1.)
[83] 'ausschlaggeben über kampf und sieg wird in ihre hand gelegt.' (*Mythologie*, p. 392.)
[84] *Schr.*, vi. 23.

however, that there was no overall regulation of choice common to all Germanic consciousness, or to suggest that the authors of *Helreidh* and the *Hákonarmál* produced such divergent accounts because they knew little and cared less about each others' portrayals of the power hierarchy: such was not the way of Romantic scholarship. Some unifying idea needed to be found which would emcompass all differences, resolve all variance, and in relation to which both *Helreidh* and *Hákonarmál* could be seen to be making a statement.

For Frauer the exact dimensions of the valkyries' choice was bound up with the wider question of fate and free will in Germanic thought. That at different times different principles of choice appeared to be operating was undeniable. To Frauer this suggested not that a common Germanic consciousness in such matters did not exist, but that it was of a profoundly complex and problematical nature. The result of this 'Germanic consciousness, divided and fragmented at its deepest level'[85] was a system of choice and its implementation which instead of running in a straight, hierarchical line took the form of a series of counterbalancing circles, often competing and not infrequently conflicting.[86]

In the philosophical tussle between Brünnhilde and her father in Act II of *Die Walküre* and its resolution in Act III we can observe Frauer's reasoning. Odin/Wotan as war-god governs the fate of heroes and the outcome of battle. He may exercise his power by personal intervention, or through the agency of his valkyries. In this respect the valkyries have no independent choice, but are simply executrices of Wotan's will. According to Frauer: 'As fulfillers of Odin's will, as his maidens and servants, the valkyries are really only subordinate beings.'[87] 'They are nothing but Odin himself as he intervenes in the lives of the heroes.'[88] Such exactly is Brünnhilde's condition as she appears at the opening of *Die*

---

[85] 'im tiefsten Grunde gespaltene und gebrochene Bewusstsein des Germanen' (*Die Walkyrien*, p. 41).

[86] *Die Walkyrien*, pp. 39–40.

[87] 'Als Vollstreckerinnen von Ôdhins Willen, als seine Mädchen und Dienerinnen sind die Walkyrien eigentlich nur untergeordnete Wesen.' (*Die Walkyrien*, p. 44.)

[88] 'Sie sind nichts Anderes, als Odhin selbst in seinem Eingreifen in das Leben der Helden.' (*Die Walkyrien*, p. 43.)

Walküre, Act II, a subordinate being with no independent existence, a mere extension of her father's will:

> *Wotan*
> mit mir nur rat' ich,
> red' ich zu dir.[89]
>
> *Wotan*
> I consult just myself
> in speaking to you.

Brünnhilde herself meekly echoes her father's sentiment:

> Zu Wotans Willen sprichst du,
> sag'st du mir, was du willst:
>   wer—bin ich,
>   wär' ich dein Wille nicht?[90]
>
> You speak to Wotan's will
> if you tell me what you wish:
>   who am I,
>   if not your will?

Even Fricka can find no quarrel:

>   Deinen Willen
> vollbringt sie allein.[91]
>
>   She fulfils
> your will alone.

Yet operating against this principle of Wotan-choice and valkyrie dependency Frauer perceived a second, contrary principle of free-acting valkyries. Troubled Germanic consciousness demanded an alternative system:

But here again Germanic man cannot avoid restoring a kind of independence and a parallel role to those it has declared subordinate, thereby bringing about a glaring contradiction. In the way they influence particular situations the valkyries are usually represented as though they were acting freely and independently, with the result that they automatically push their lord and master Odin into the background and completely take his place. We shall

---

[89] *Schr.*, vi. 37.
[90] *Schr.*, vi. 37.
[91] *Schr.*, vi. 34.

even cite an example (from the heroic sagas) where the decision of a valkyrie directly conflicts with Odin's will.[92]

This is the Brünnhilde Frauer knows and as she appears in Wagner's drama at the moment of her conversion. To attain world-redemptive compassionate womanhood Brünnhilde must emancipate herself from Wotan's will; the road from divinity to humanity lies through disobedience:

Wotan
Durch meinen Willen
warst du allein:
gegen ihn doch hast du gewollt . . .[93]

Wotan
Through my will only
you existed;
yet you have chosen against it . . .

For Frauer the differences between god and valkyrie add up to a system of starkly opposing principles of choice. Wagner presents them, in more reconciliatory light, as complementary ones. Brünnhilde has challenged Wotan's will; but will is not the only operative factor in choice. As well as the dynamic force of the will there is the affective power of desires, from which spring wishes. Wotan, we shall shortly see, is the dynamic force and the ruling will of the universe; but he is none the less equally the god of wishes. Oski or Wunsch, one of the names he goes under, means 'wish'. The valkyries too are known as 'oskmeyar': 'wish-maidens'—'I imagine,' says Grimm, 'because they are in Odin's service and Odin is called "Oski" or "Wunsc".'[94] For

---

[92] 'Aber auch hier kann der Germane nicht umhin, dem als untergeordnet Ausgesprochenen wieder eine Art von Selbstständigkeit und Nebenordnung zu verleihen, und so einen schroffen Gegensatz hervorzurufen. Die Walkyrien werden bei ihrem Einwirken auf bestimmte Verhältnisse meist so dargestellt, als ob sie frei und selbstständig handelten, wodurch sie von selbst ihren Herrn und Meister Ôdhin zurückdrängen und ganz an seine Stelle treten. Wir werden (aus der Heldensage) sogar ein Beispiel anführen, wo die Entscheidung einer Walkyrie dem Willen Ôdhins geradezu widerspricht.' (*Die Walkyrien*, p. 44.)

[93] *Schr.*, vi. 72.

[94] 'Ich denke, weil sie in Odins diensten stehen und Odinn *Oski, Wunsc* heisst.' (*Mythologie*, p. 390.) Frauer gives an alternative explanation: 'weil sie von Odhin ausgewählt, angenommen sind zu seinem Dienste.' ('because they have been chosen by Odin and taken into his service.' *Die Walkyrien*, p. 1.)

Wagner they are variously 'Wunschmädchen'[95] and 'Wunsches-Mädchen',[96] wish-maidens and daughters of Wish. Brünnhilde especially is the 'creative womb of his wishes'.[97] It is the valkyries' job to carry out Wotan's wishes every bit as much as to implement his will.

In defending Siegmund, now, Brünnhilde is acting contrary to Wotan's orders, yet in accordance with his desires. Wotan, with some bitterness, admits as much:

> So tatest du,
> was so gern zu tun ich begehrt—
> doch was nicht zu tun
> die Not zwiefach mich zwang?[98]

> So you have done
> what I so yearned to do,
> but which not to do
> necessity doubly constrained me?

Brünnhilde is acting *for* Wotan in her capacity of wish-fulfiller, *against* him in her role as agent of his will. It is not her fault if the two roles are incompatible. Conflicts are there in Wagner's system, but they are not essentially between god and valkyrie, rather within the god himself: tragic conflicts between duty and inclination, between natural law and imposed order, between responsibility for a universe and love for a son. Brünnhilde's disobedience is a rescue bid, saving vital aspects of her father's nature from annihilation, undertaken

> Weil für dich im Auge
> das *eine* ich heilt,
> dem, im Zwange des *andren*
> schmerzlich entzweit,
> ratlos den Rücken du wandtest.[99]

> Because the *one* thing
> I kept in view
> which in painful estrangement,
> compelled by the *other*,

[95] *Schr.*, vi. 50.
[96] *Schr.*, vi. 51.
[97] 'Wunsches schaffender Schoss' (*Schr.*, vi. 72).
[98] *Schr.*, vi. 78.
[99] *Schr.*, vi. 77.

you turned your back on, baffled.

Choice in Wagner's drama is shared between Wotan and
Brünnhilde. Together their choices offer a complete answer
to all the needs of the situation. However, the incompatibility of their choices entails severance. From now on god
and valkyrie go their separate ways; but with her Brünnhilde
takes all Wotan's hopes and wishes and innermost thoughts.

In his Young Siegfried drama Wagner had first brought
Wotan on stage as the seigneurial Wanderer and as the enigmatic Voma, herald of Day. Once his involvement with the
god had begun Wagner found himself increasingly fascinated
by the complexities and ambiguities of the god's personality.
It was with Wotan that he came to identify more than with
any other character in the *Ring*.[100] What had started as a
Siegfried drama now became first and foremost a Wotan
drama.[101]

In the stage action of *Das Rheingold* and *Die Walküre*, but
also in the great debates these two works contain, Wagner
completed his many-tiered portrait of the god. Scenes such as
Wotan's confrontation with Fricka in Act II of *Die Walküre*
teem with allusions to the god's vastly varied characteristics.
Indeed, so philosophical did that particular scene become that
Wagner omitted several passages of the original poem when
he came to compose the text, in order not to hold up the
dramatic action unduly. They were, however, printed in the
Sämtliche Schriften.[102]

The task Wagner set himself with his Wotan portrait
reflects the deep influence of Romantic scholarship on his
thinking. For as Frauer had sought a unifying idea to accommodate all the variations in the sources on matters of choice,
necessity, and free will; or as von der Hagen conversely had
found in every point of similarity a chain of unity which
could embrace Siegfried, Hagen, Dietrich, and Baldur:[103] so

---

[100] Letter to August Röckel, 25 Jan. 1854, *Dokumente*, p. 93: 'Sieh Dir ihn [Wotan]
recht an! er gleicht *uns* auf's Haar . . .' ('Take a good look at Wotan! He resembles *us*
to a T . . .').
[101] Golther, *Grundlagen*, p. 90.
[102] vi. 25–31.
[103] *Die Nibelungen*, pp. 78–85.

Wagner took material spanning several centuries, from many lands and a thoroughly heterogeneous array of sources, and determined to weld it into a Wotan portrait of psychological verisimilitude. The quantity and diversity of source material were vast. Grimm's guide to the god in the chapter 'Wuotan' of his *Mythologie* remained Wagner's chief aide, bringing his philological knowledge to bear on the god's many names and enriching what Wagner found in the Eddas and other Scandinavian sources with material from medieval literature and folk saga. In addition, here more than anywhere else in the *Ring* we shall have grounds to consider Wilhelm Müller's *Altdeutsche Religion* as a possible contributor.

'Oski' or 'Wunsch' was one of the kindlier of Wotan's identities, and particularly dear to Jacob Grimm, who devotes a substantial section of his 'Wuotan' chapter to this benign face of the god. Grimm draws upon more than two score quotations from medieval German poetry to demonstrate that as 'Wish' Wotan represented for the ancient Germans the 'epitome of all happiness and bliss, the fulfilment of all gifts.'[104] It is this mild aspect of his divine father that Siegmund invokes in Act I of *Die Walküre* when to Sieglinde he says:

> Unheil wende
> der Wunsch von dir![105]
>
> May Wish turn
> all trouble from you!

—lines which within the context of the drama and the twins' eventual fate have their own special pathos.

If 'Wish' suggests serenity after the storm, Wotan is no less present in the storm itself. According to Grimm, the semantic origin of the name Wotan, like the word 'Wut' ('rage'), is 'mens [mind], ingenium [spirit], then vehemence and wildness'.[106] Turbulence and tempestuousness are part of Wotan's nature. In Act III of the Young Siegfried drama the Wala calls

---

[104] 'inbegriff von heil und seligkeit, die erfüllung aller gaben.' (*Mythologie*, p. 126).

[105] *Schr.*, vi. 4.

[106] 'mens, ingenium, dann ungestüm und wildheit' (*Mythologie*, p. 120).

her disturber 'unruly hothead';[107] Fricka in Act II of *Die Walk-*
üre terms her consort 'raging god' and 'raving'.[108]
Grimm writes that Wotan 'approaches through the air in a
gale'.[109] The god who at the opening of Act II of *Siegfried*
draws near in the rustling breeze arrives in Act III of *Die*
Walküre chasing on the wings of the storm:

> *Waltraute*
> Nächtig zieht es
> von Norden heran.
> *Ortlinde*
> Wütend steuert
> hierher der Sturm.
> *Die Walküren*
> Wild wiehert
> Walvaters Ross,
> schrecklich schnaubt es daher![110]

> *Waltraute*
> Dark as night
> it draws from the north.
> *Ortlinde*
> The storm is furiously
> steering this way.
> *The valkyries*
> Walvater's horse
> is whinnying wildly,
> fearfully snorting from far!

Excitement mounts on stage and in the orchestra until finally
the raging Heervater is at hand: 'The summit of the rock is
surrounded by black clouds; a terrible storm is raging in the
background; a fiery light illuminates the pine forest at the
side. Between thunderclaps Wotan's call can be heard.'[111]
The arrival of Wotan, whose name 'Yggr' signifies 'Ter-
ror',[112] fills Brünnhilde with just apprehension:

[107] 'störrischer Wilder' (*Schr.*, vi. 155).
[108] 'rasender Gott', 'Wütender' (*Schr.*, vi. 30, 26).
[109] 'durch die lüfte im sturm naht' (*Mythologie*, p. 135).
[110] *Schr.*, vi. 65.
[111] 'Die Felsenhöhe ist von schwarzen Gewitterwolken umlagert; furchtbarer
Sturm braust aus dem Hintergrunde daher: ein feuriger Schein erhellt den Tan-
nenwald zur Seite. Zwischen dem Donner hört man *Wotans* Ruf.' (*Schr.*, vi. 70.)
[112] Grimm, *Mythologie*, p. 132.

> Der wilde Jäger,
> der wütend mich jagt,
> er naht, er naht von Norden![113]

> The Wild Huntsman
> in hot pursuit
> draws near, draws near from the north!

The Wild Huntsman of popular superstition was, as Grimm portrays him, Wotan's most dangerous surviving manifestation. Grimm has much to say and many reports to record of the Wild Huntsman and his 'wütendes Heer', riding eerily through forest and sky at night with tremendous din of baying hounds and hunting calls. According to folk saga, it is better to stay safely indoors when the Wild Huntsman and his troops pass by. Brave spirits abroad may find themselves engaged in a trial of strength with Wod. The incautious and the impudent are swiftly mown down.[114] One or two of the tales Jacob Grimm mentions on his *Mythologie* were known to Wagner in their entirety from the brothers' *Deutsche Sagen*.[115] Such is the fear-inspiring approach of Wotan now as he pursues his disobedient daughter, for as Jacob Grimm remarks: 'Even Wotan's "wütendes Heer", what else is it but an interpretation of the storm wind howling through the air?'[116]

The leader of the ghostly army of latter-day folk belief was in Eddic times the god of war. Grimm singles out three appropriate names denoting Wotan's martial role—Herfadir, Valfadir and Sigfödr[117]—which Wagner duly added to his range in *Die Walküre* as Heervater,[118] Walvater,[119] and Siegvater.[120] Wotan is lord of armies, lord over life and death in battle and of the slain, lord of victory.

The conduct and outcome of war are only part of Wotan's concern as war-god. His responsibilities extend to the fur-

[113] *Schr.*, vi. 64.
[114] Grimm, *Mythologie*, pp. 870–902.
[115] e.g. Nos. 48, 172.
[116] 'Selbst Wuotans wütendes heer was ist es anderes als eine deutung des durch die luft heulenden sturmwindes?' (*Mythologie*, p. 599.)
[117] *Mythologie*, pp. 121–2.
[118] e.g. *Schr.*, vi. 35.
[119] e.g. *Schr.*, vi. 49.
[120] e.g. *Schr.*, vi. 45.

therance and promotion of warfare. He engenders a climate favourable to hostilities; in his own words:

> ... wo kühn Kräfte sich regen,
> da gewähr' ich offen den Krieg.[121]

> ... where forces are stirring to action
> I openly license the war.

If the forces do not rouse themselves of their own accord Wotan helps them, stirring up strife and discord among the heroes and encouraging ambitions. Fricka accuses:

> durch deinen Stachel
> streben sie auf:
> du—reizest sie einzig.[122]

> spurred on by you
> they strain at the yoke:
> you alone incite them.

Incitement to battle is one of the tasks Heervater may delegate to the valkyries. He reminds the disgraced Brünnhilde:

> Helden-Reizerin
> warst du mir ...[123]

> Inciter of heroes
> you were for me ...

The trouble-making side of Wotan's war-god activity is not one which can be laid at Grimm's door. Wilhelm Müller, however, remarks: 'Every hostility, every war is promoted by Odin.'[124]

Wanderer, god of turbulence and tumult, stirrer of air and of heroes: all point to a common factor in Wotan's being, that of movement. At the very beginning of the god's name, before even the noun 'Wut', Grimm places the verb 'waten ... which corresponds literally with the Latin "vadere" [walk] and signifies "meare, transmeare, cum impetum ferri" [go, traverse, rush wildly].'[125]

---

[121] *Schr.*, vi. 29.

[122] *Schr.*, vi. 32.

[123] *Schr.*, vi. 73.

[124] 'Jede feindschaft, jeder krieg wird duch Odhinn erregt.' (*Altdeutsche Religion*, p. 197.)

[125] 'waten ... welches buchstäblich dem lat. vadere entspricht, und meare, transmeare bedeutet, cum impetum ferri.' (*Mythologie*, p. 120.)

Grimm's definition gives us the Wanderer but also, more broadly, Wotan as the dynamic force of the universe. Probably it was from this idea of the god as perpetual motion that Wagner developed his image of a god of evolution and change. He is present in the constant flux of forces:

> wo Kräfte zeugen und kreisen,
> zieh' ich meines Wirkens Kreis . . .[126]
>
> where milling forces multiply
> I mark out my sphere of action . . .

He is the god of renewal, of the 'eternally young'.[127] He governs the ebb and flow of fortunes and the cosmic cycle, a role which dates back to *Die Wibelungen*, where Wagner set out his thoughts on the god in a rather involved passage:

The primeval struggle is therefore carried on by us, and its variable success is exactly the same as the constantly recurring alternation of day and night, summer and winter—and finally, within the human race itself, which progresses ever onward from life to death, from victory to defeat, from joy to sorrow . . . The essence of this perpetual motion—of life in other words—eventually found its expression in 'Wuotan' (Zeus), being the principal god, the father and pervader of the universe . . .[128]

The passage quoted above reveals another paradoxical aspect of Wotan: the god of perpetual motion is constantly present. As 'father and pervader of the universe' Wotan is the all-pervasive spirit, the life-force. Grimm again was responsible, for both teasing out this definition of Wotan's role from the etymology of the name and, in so doing, demonstrating the association with movement; for he derives the 'Wut' element—normally 'rage'—in 'Wotan' from the verb of movement 'waten' ('walk' or 'go') and gives as its prime meaning, before 'vehemence and wildness', 'actually "mens,

---

[126] *Schr.*, vi. 31.
[127] 'ewig Jungen' (*Schr.*, vi. 157).
[128] 'Der uralte Kampf wird daher von uns fortgesetzt, und sein wechselvoller Erfolg ist gerade derselbe, wie der beständig wiederkehrende Wechsel des Tages und der Nacht, des Sommers und des Winters,—endlich des menschlichen Geschlechtes selbst, welches von Leben zu Tod, von Sieg zu Niederlage, von Freude zu Leid sich fort und fort bewegt . . . Der Inbegriff dieser ewigen Bewegung, also des Lebens, fand endlich selbst im '*Wuotan*' (Zeus), als dem obersten Gotte, dem Vater und Durchdringer des Alls, seinen Ausdruck . . .' (*Schr.*, ii. 132).

ingenium" [mind, spirit]'.[129] Wotan is consequently 'the all-
powerful, all-pervasive being . . . the god of spirit.'[130]

The primeval energy which erupts so destructively as the
Wild Huntsman can, with organization and control, become
creative. The storm-clad god of *Die Walküre*, Act III, is in Act
III of *Siegfried* himself named 'Subduer of Storms'.[131] First
among Wotan's attributes Grimm puts 'the all-pervasive creative and shaping force, bestowing form and beauty on men
and all things.'[132] Wagner gives a sample of Wotan's power
over matter in Loge, the restless fire-spirit. Loge has obtained
from Wotan form and shape; at the close of *Die Walküre* the
hand that bound him releases Loge into his free element once
more:

> Loge, hör!
> lausche hieher!
> Wie zuerst ich dich fand
> als feurige Glut,
> wie dann einst du mir schwandest
> als schweifende Lohe:
> wie ich dich band,
> bann' ich dich heut'!
> Herauf, wabernde Lohe,
> umlodre mir feurig den Fels![133]

> Hark, Loge,
> listen this way!
> As first I found you,
> a fiery glow;
> as once you ran from me,
> roving fire;
> as you were when I caught you
> I conjure you now!
> Arise, flickering flames!
> Blaze fiercely round the rock!

Animating spirit of the universe, bestower of form, giver of

---

[129] 'eigentlich mens, ingenium' (*Mythologie*, p. 120). See n. 106 above.
[130] 'das allmächtige, alldurchdringende Wesen . . . die geistige Gottheit.'
(Grimm, *Mythologie*, p. 120.)
[131] 'Stürmebezwinger': *Schr.*, vi. 154.
[132] 'die alldurchdringende schaffende und bildende kraft, der den menschen und
allen dingen gestalt wie schönheit verleiht.' (*Mythologie*, p. 121.)
[133] *Schr.*, vi. 84.

life: the Edda speaks of Wotan as 'Allvater', 'universal
father'. In the *Die Wibelungen* passage on Wotan we quoted
earlier Wagner had termed the god 'Vater . . . des Alls':
'father of the universe'. The name Allvater itself is solidly
embedded and solemnly enshrined in the closing lines of
Siegfrieds Tod:

> Nur einer herrsche:
> Allvater! Herrlicher du![134]
>
> One only shall govern:
> Allvater the glorious! You!

Having sired the universe, Wotan now rules it. The system
in force in 1848–9, in both the Nibelung drama and *Die
Wibelungen*, is staunchly patriarchal. Generally accepted as
'the highest and chief among the gods',[135] Wotan was conceived of by Wagner as exercising a father's authority over all
creation and, indeed, over his fellow gods: to quote from *Die
Wibelungen* again, Wotan 'was by his very nature bound to be
considered the supreme god, and as such was also bound to
occupy the role of father to the other gods . . .'[136]

If there were only the evidence of *Die Walküre* to go on we
might be in some doubt as to whether patriarchy was still
holding its ground in the *Ring* of 1852, or whether indeed
matriarchy had taken over. But in *Das Rheingold* Wotan's
authority over his fellow gods is convincing enough. When
Thor seeks to settle the matter of the giants' claim with his
hammer Wotan intervenes to uphold law and order.[137] The
deciding voice in council and the guardian of justice, Wotan
is confirmed as Head of State.

With the establishment of Wotan's political authority
Wagner is leaving the sphere of Grimm's influence and,
perhaps, entering that of Wilhelm Müller, who writes in his
Altdeutsche Religion:

Since the guiding and ordering of human affairs, and their joy and
sorrow, lay in the hands of the wise sky-god [Odin], it was particu-

---

[134] *Schr.*, ii. 227.

[135] 'die höchste und oberste gottheit' (Grimm, *Mythologie*, p. 120).

[136] 'musste . . . seinem Wesen nach als höchster Gott gelten, als solcher auch die
Stellung eines Vaters zu den übrigen Gottheiten einnehmen . . .'(*Schr.*, ii. 132).

[137] *Schr.*, v. 221–2.

larly Odin who became the state god in the north, and his cult
thereby obtained a high political significance which the veneration
of other gods lacked. All the main institutions of the state and all
public activities which influence its destiny appear to have been
placed under his special protection . . .

In the *Ynglinga saga* (Chapter 8) Snorri represents Odin as law-
giver generally, and indeed it appears that not only the government
of the state stood under his protection but also all legal matters.[138]

The emblem of Wotan's political and judicial authority is
his spear, seen in action at its most incisive when Thor
attempts to short-change the giants:

> *Wotan*
> (*seinen Speer zwischen den*
> *Streitenden ausstreckend*)
> Halt, du Wilder!
> Nichts durch Gewalt!
> Verträge schützt
> meines Speeres Schaft . . .[139]

> *Wotan*
> (*holding out his spear between*
> *the disputants*)
> Stop, you firebrand!
> Nothing by force!
> The shaft of my spear
> safeguards contracts . . .

The spear as an adjunct of Wotan's war-god role was familiar
to the mythologies of Wagner's day. Mostly their material
was based on the *Heimskringla* and other sagas. Casting a
spear opens hostilities; a spear-wound marks heroes out for
Wotan.[140] As symbol of Wotan's civil authority, however,

---

[138] 'Indem nun in der hand des weisen himmelsgottes die lenkung und ordnung
der angelegenheiten der menschen und ihr wohl und wehe steht, wird Odhinn
insbesondere der staatsgott des nordens, und sein kultus bekam dadurch eine hohe
politische bedeutung, welche der verehrung anderer götter entging. Alle haup-
teinrichtungen des staates, alle öffentlichen handlungen, welche auf das geschick
desselben einfluss haben, scheinen unter seinen besondern schutz gestellt zu sein . . .
Snorri stellt in der Ynglinga-saga (c. 8) Odhinn überhaupt als gesetzgeher hin, und
es scheint allerdings, dass nicht nur die regierung des staates, sondern auch alle
rechtlichen handlungen unter seinem schutz standen.' (pp. 191–2.)

[139] *Schr.*, v. 222.

[140] Grimm, *Mythologie*, p. 134; Müller, *Altdeutsche Religion*, pp. 197–8.

the spear did not arouse the same attention. One looks in vain for reference to the civilian role of Wotan's spear in Grimm, either in the *Mythologie* or in the *Deutsche Rechtsalterthümer*. Any guidance Wagner got can only have come from Wilhelm Müller, who writes in his *Altdeutsche Religion*:

> We may also quite rightly surmise that under the protection of the god [Odin] were particularly such legal matters in which symbols were formerly used which we also find associated with him. Thus the spear Gungnir was Odin's chief weapon . . . Among the Franks it also symbolically represents the highest authority.[141]

If Wagner had a source for the part played by Wotan's spear in *Das Rheingold*, therefore, Müller is the obvious candidate. Elsewhere in his *Altdeutsche Religion*, of course, Müller draws heavily on Grimm; indeed, many facets of Wotan's portrait that Wagner derived initially and ultimately from Jacob Grimm are mentioned by Müller too. Supposing that Wagner was engaged in some reading or rereading for his Nibelungen drama currently in Zurich, we might speculate that it included Müller's *Altdeutsche Religion* and that this acted as Wagner's guide not only in matters where it diverges from Grimm, such as the spear, but also where it converges, reminding the composer of what he had once found when grappling with Grimm while taking the waters at Teplitz (see Chapter 1).

Wotan's spear proved a visual image so potent that once he had introduced it into his drama Wagner continued to exploit its symbolism. By its power Loge is summoned at the close of *Die Walküre*.[142] In defence of established order Wotan wields it against the wayward Siegmund's sword.[143] In the counterpart scene to this, introduced subsequently by Wagner when he revised his Young Siegfried drama, the world-dominating shaft of Wotan's spear is shattered by Siegfried's sword of independence.[144] After the ring itself

---

[141] 'Wir dürfen auch mit recht vermuten, dass namentlich solche rechtliche handlungen unter dem schutze des gottes stunden, bei welchen ehemals symbole angewandt wurden, die wir bei ihm wieder finden. So war der speer, Gungnir genannt, Odhins hauptwaffe . . . bei den Franken bezeichnete derselbe auch symbolisch die höchste gewalt.' (p. 193.)

[142] *Schr.*, vi. 84.

[143] *Schr.*, vi. 57.

[144] *Schr.*, vi. 163.

Wotan's spear became the most important symbol in the power structure of Wagner's Nibelungen drama, so much so that Wagner cleared whole areas in the Norns' Prologue to *Götterdämmerung* and the knowledge-contest scene of *Siegfried*, Act I, to make room for a description of the origin and properties of Wotan's spear. Drawing together from his primary sources the motifs of the world tree, Wotan's invention of runes and their inscription on weapons, Wagner produced the following account:

> Aus der Welt-Esche
> weihlichstem Aste
> schuf er sich einen Schaft:
> dorrt der Stamm,
> nie verdirbt doch der Speer;
> mit seiner Spitze
> sperrt Wotan die Welt.
> Heil'ger Verträge
> Treue-Runen
> sind in den Schaft geschnitten:
> den Haft der Welt
> hält in der Hand,
> wer den Speer führt,
> den Wotans Faust umspannt.
> Ihm neigte sich
> der Nibelungen Heer;
> der Riesen Gezücht
> zähmte sein Rat:
> ewig gehorchen sie alle
> des Speeres starkem Herrn.[145]

> From the World Ash's
> most worshipful bough
> he shaped for himself a shaft.
> The stem may fade
> but the spear will not fail;
> with its point
> Wotan holds fast the world.
> Constancy-runes
> of solemn contracts
> are carved into the shaft;
> he bears in his hand

[145] *Schr.*, vi. 104.

the bond of the world,
who carries the speer
which Wotan's fist encompasses.
The Nibelung host
bowed to him;
his counsel tamed
the giant tribe:
they all submit for ever
to the spear's mighty lord.

Immutable treaties engraved in potent runes on Wotan's spear give the god power and dominion over the rest of the world's citizens. The spear is more than just an emblem of Wotan's authority: his power is vested in it.

Wotan's readiness to protect the beleaguered giants from Thor's hammer in *Das Rheingold*, Scene ii, is now explained. It is not just a gentlemanly gesture, nor is Wotan moved by purely dispassionate concern for justice. Least of all is he swayed by the immediate or short-term interest of the gods, to whom the elimination of the giants and their claim would be most welcome. His spear intervenes to honour the giants' contract because it is by contracts that Wotan rules, and any breach of contract jeopardizes his dominion. The god who rules by contracts is bound by them too. Only a short while previously Fasolt has thought fit to remind the temporizing Wotan:

Lichtsohn du,
leicht gefügter,
hör' und hüte dich:
Verträgen halte Treu'!
Was du bist,
bist du nur durch Verträge:
bedungen ist,
wohl bedacht deine Macht.[146]

Son of light,
airy substance,
listen and look to yourself:
keep your faith with contracts!
What you are,

146 *Schr.*, v. 219.

you are through contracts alone:
your power's been settled,
given much consideration.

Das Rheingold finds Wotan as much in command as in the Nibelungen drama of 1848. Yet between the two there has been an internal revolution. By sleight of hand, so it seems, the god who in *Die Wibelungen* reigned supreme 'by his very nature' depends in the *Ring* for his power on the runic treaties carved into the shaft of his spear. It suited Wagner's current political philosophy that the soon-to-be-discredited race of gods should be seen as the representatives of a rigid system of contractual law. But the god who finds himself obliged to maintain the fixed order and uphold the stability of the universe is himself the god of mobility and change, of evolution and upheaval. This is the paradox at the heart of Wotan's dispute with Fricka in *Die Walküre*, Act II, and at the root of the god's inner division, out of which Brünnhilde attempts to rescue him:

> *Fricka*
> Wohin rennst du,
> rasender Gott,
> reissest die Schöpfung du ein,
> der selbst das Gesetz du gabst?
> *Wotan*
> Des Urgesetzes
> walt' ich vor allem:
> wo Kräfte zeugen und kreisen,
> zieh' ich meines Wirkens Kreis . . .[147]

> *Fricka*
> Where are you rushing to,
> reckless god?
> Are you destroying creation,
> whose laws you established yourself?
> *Wotan*
> I exercise foremost
> this ancient law:
> where the milling forces multiply
> I mark out my sphere of action . . .

The duel between existing order and innovation, externally

---

[147] *Schr.*, vi. 30–1.

represented by Fricka's demands and the claims of Siegmund, has its origin in Wotan's complex personality.

In the same way that Frauer reconstructed pre-Christian
Germanic consciousness, Wagner, pursuing the Romantic
principles of the intrinsic unity underlying all manifestations
and the common origin of every report, blended his Wotan-
portrait from the wide-ranging sources he had come across in
his researches. As a portrait it works; any theatre-goer will
confirm that Wotan is one of the most convincing characters
in the *Ring* despite the inconsistencies and volatilities of
mood which result from such a method of construction. But
the resulting god, like Frauer's 'Germanic man', is one torn
by inner strife and at war with himself. And just as the
'divided and fragmented consciousness' of Frauer's Teuton
'can only attain a glimmer of reconciliation by itself destroy-
ing once more the divine powers it had itself created',[148] so
Wagner's Wotan finds inner reconciliation only in his own
annihilation:

> dem ewig Jungen
> weicht in Wonne der Gott.[149]
>
> to eternal youth
> the god yields with joy.

Wagner's other gods, Donner and Froh, have barely progressed from their 1848 level. They remain psychologically
underdeveloped, individualized only by superficial features,
conventional theatre figures. Much the same is true of Freia,
whom Wagner introduced as symbol of love following
Grimm's definition of her as 'goddess of love and fertility', as
distinct from 'the divine mother and guardian of marriage',
Fricka.[150] It was a hard-won distinction, for the demarcation-
lines between the two goddesses were very fluid in the
sources and Grimm spends several pages of his chapter on
goddesses disentangling them.[151] Germanic religion, Grimm

---

[148] 'weiss sich einen Schimmer von Versöhnung nur zu erringen, indem es die
von ihm selbst geschaffenen göttlichen Mächte selbst wieder zertrümmert' (*Die
Walkyrien*, p. 41).

[149] *Schr.*, vi. 157.

[150] 'göttin der liebe und fruchtbarkeit', 'die göttliche mutter und vorsteherin der
ehe' (Grimm, *Mythologie*, p. 280).

[151] *Mythologie*, pp. 276–9.

discovered, had altogether a capacity for producing goddesses of similar, chiefly domestic competences, including such
figures of contemporary folk belief as Bertha and Frau Holle
or Holda. Holda and Freia are both listed among the 'goddesses who have evolved from moral qualities . . . Holda the
gracious ["hold"], Freia the fair or joyful ["froh"] . . .'[152] — a
passage which, amended, was no doubt the inspiration
behind Fasolt's:

> Freia, die holde,
> Holda, die freie . . .[153]
>
> Freia the gracious,
> Holda, the free . . .

Perhaps it was Grimm's reference to Freia as a fertility-
goddess which prompted Wagner to identify her with Idunn,
guardian of the apples of youth in the Edda. Idunn is another
victim of Loge's machinations. Like Freia, she is bartered to
the giants; when she goes her apples go with her and the gods
begin to age.[154] As a late entry in *Das Rheingold* Wagner gave
Idunn's apples into Freia's keeping, both to increase the number of the gods' dilemmas with her absence and to allow
some chilling stage effects. As Freia recedes towards Riesen-
heim: 'A pale mist fills the stage, growing denser; in it the
gods assume an increasingly wan and aged appearance; all
stand looking anxiously and expectantly at Wotan, who is
gazing down reflectively at the ground.'[155] Meanwhile Loge
cheerily comments:

> Jetzt fand ich's: hört, was euch fehlt!
> Von Freias Frucht
> genosset ihr heute noch nicht:
> die gold'nen Äpfel
> in ihrem Garten,
> sie machten euch tüchtig und jung,
> ass't ihr sie jeden Tag.

---

[152] 'aus sittlichen begriffen hervorgegangenen göttinnen . . . Holda die holde . . .
Frouwa, Freja die schöne oder frohe . . .' (Grimm, *Mythologie*, pp. 842–3).
[153] *Schr.*, v. 218.
[154] Simrock, *Edda*, pp. 259, 291.
[155] 'Ein fahler Nebel erfüllt mit wachsender Dichtheit die Bühne; in ihm erhalten
die Götter ein zunehmend bleiches und ältliches Aussehen; alle stehen bang und
erwartungsvoll auf Wotan blickend, der sinnend die Augen an den Boden heftet.'
(*Schr.*, v. 231.)

Des Gartens Pflegerin
ist nun verpfändet:
an den Aesten darbt
und dorrt das Obst:
bald fällt faul es herab.

. . .

Ohne die Äpfel
alt und grau,
greis und grämlich,
welkend zum Spott aller Welt,
erstirbt der Götter Stamm.[156]

Listen! I've found what you're lacking:
   today you've not eaten
any of Freia's fruit;
   the golden apples
   in her garden
caused you to keep strong and young
if you ate them every day.
   The garden's mistress
   is mortgaged now;
   the crop withers
   and wastes on the boughs:
soon the fruit will rot and fall.

. . . .

Without the apples,
   old and grey,
   aged and anxious,
wilting, the butt of the world,
the line of the gods will lapse.

The peculiar air of paralysing calamity which grips the scene, not least in the music, is the first evidence of the poem *Hrafnagaldr Odhins* in the *Ring*. Simrock had included *Hrafnagaldr Odhins* in his 1851 *Edda* despite the poem's dubious authenticity. Autumnal in mood, heavy and doom-laden, the poem is obscure even by Eddic standards and in need of the kind of interpretative commentary Simrock gives it in his notes to clarify the narrative progression.

Simrock sees *Hrafnagaldr Odhins* as a prelude to the *Vegtamskvidha*, which in his edition follows it. It is an Idunn

[156] *Schr.*, v. 232.

poem: the goddess has sunk down beneath the earth; the world tree wilts and all creation falters. Strengths seize and powers fail. Perplexity and dismay overcome the gods:

> 2. Die Asen ahnten
>   Übles Verhängniss,
>   Verwirrt von widriger
>   Wesen Zeichen.
>
>   .   .   .   .   •
>
> 23. Die Kräfte ermatten,
>   Ermüden die Arme,
>   Schwindelnd wankt
>   Der weisse Schwertgott.
>   Es ebbt der Strom
>   Der eisigen Luft
>   Und betäubt die Sinne
>   Der ganzen Versammling.[157]
>
> 2. The Aesir suspected
>   An evil destiny,
>   Dazed by omens
>   Of adverse beings.
>
>   .   .   .
>
> 23. Powers fade,
>   Arms fail,
>   The white sword-god
>   Wavers, irresolute.
>   The flow ebbs
>   Of icy air
>   And dulls the senses
>   Of all the assembly.

The spell is broken only when Odin rouses himself to decisive action and sets forth—in the *Vegtamskvidha*, Simrock suggests—on a journey to the underworld.

In the Edda Wotan descends to Nibelheim to seek out the Wala; in *Das Rheingold* the Wala rises to seek out the god. In the Young Siegfried drama, true to the *Vegtamskvidha* original, she had risen like a corpse from the grave. In *Das Rheingold* she sports a new name, Erda, and appears in

---

[157] Simrock, *Edda*, pp. 33, 36.

rejuvinated form as the earth goddess: 'She is of noble stature, surrounded by a mass of black hair.'[158] While there was a fair degree of evidence for earth-mother cults in Germanic territory Grimm could find no single, well-documented earth goddess and had to fill out the concept with examples from classical mythology: 'Gaea, Tellus, Terra reappear as our goddesses Fiörgyn, Iord, and Rindr.'[159] The classical models had the advantage that they produced offspring such as the Fates and Prometheus;[160] Wagner's Erda is mother of Brünnhilde and the norns.[161] An earth goddess in the classical mould, Wagner must have decided, provided a more convincing arch-mother figure, and a more alluring prospect for Wotan, than did the *Vegta—skvidha* corpse.

Introducing the giants into *Das Rheingold* signalled a retreat from the mass politics of the *Mythus* in favour of a simpler, more personal approach to the species. In some ways it was a return to the sources, where the ransom demand had been very much a family affair. Fafner, in fact, Wagner was able to take direct from the saga: his name, his greed, his parricidal tendencies, his conversion to hoard-guarding dragon, and his mean temperament. Fafner delivers the first lesson on the ring's power to destroy the sacred bonds of kinship. To do this he needs a relative to kill, and since his brother in the sources—the smith, Siegfried's foster-father—was already accounted for otherwise, Wagner supplied Fafner with a new brother, Fasolt.

Wagner had come across Fasolt both in the *Thidreks saga*, where he is not a giant, and in the *Heldenbuch* poem *Ecken Ausfahrt*, where he is. As a giant Fasolt appears to have been among the better-known and better-loved of the race; Müller devotes a relatively large slot to him in his chapter on giants,[162] and Grimm mentions him several times in the

---

[158] 'Sie ist von edler Gestalt, weithin von schwarzem Haare umwallt.' (*Schr.*, v. 261.)

[159] 'Gaea, Tellus, Terra kehren in unsern göttinnen Fiörgyn, Iord und Rindr wieder.' (*Mythologie*, p. 319.)

[160] Prometheus' mother was Themis or Gaea, according to Aeschylus. Themis was also mother of the Parcae.

[161] *Schr.*, vi. 38, 154; v. 261-2.

[162] *Altdeutsche Religion*, pp. 318-19.

course of his *Mythologie*.[163] According to Grimm the '-olt'
element in Fasolt's name was a popular ending for the names
of giants in medieval literature.[164] More likely to interest
Wagner was the other end of the giant's name, which
alliterated so gratifyingly with 'Fafner'.

In Wagner's first sketch for *Das Rheingold* the builders of
Walhall are weather-giants, Windfahrer and Reiffrost.[165]
From Müller in particular Wagner might have known Fasolt
as a storm-giant, and he was therefore able to salvage some of
his original intention when he amended the giants' names.

One particular characteristic of Wagner's Fasolt came
ready-made in the sources, in a passage from *Ecken Ausfahrt*
which Wilhelm Grimm singles out for particular mention in
his *Deutsche Heldensage*:

> Drey Helden sassen in einen Sal . . .
> Das ein das was sich Herr Fasolt
> Dem waren die schönen Frawen hold.[166]
>
> Three heroes sat within a hall . . .
> Herr Fasolt was among them there,
> He was a man for the ladies fair.

Fasolt's longing for Freia in *Das Rheingold* contrasts pleas-
ingly with the calculating shrewdness of his brother Fafner.

In other respects, too, Wagner used his second representative of the race to give a broader perspective and provide
contrast. If brother Fafner is the arch-typical wicked giant,
Fasolt is one of the simple kind.

> Ein dummer Riese
> rät dir das . . .
>
> A simple giant
> is saying this . . .

—so Fasolt concludes his lecture on honesty to Wotan.[167]
Massive, naïve, rough-hewn, Fasolt is nevertheless upright
and trustworthy and better able to speak for moral integrity

---

[163] e.g. pp. 524, 602.
[164] *Mythologie*, p. 494.
[165] *Skizzen*, p. 203.
[166] p. 214.
[167] *Schr.*, v. 219.

than anyone else there present. He represents the old and dying order of giants on which Grimm reports so favourably in the *Mythologie*: 'They represent a race which has been eclipsed or which is in decline, whose strength was accompanied by the innocence and wisdom of ancient times: more an impersonal and innate wisdom than a personally acquired understanding.'[168]

Descended of giants in the sources, numbered among the aesir, Loge is termed by Russwurn 'the profoundest puzzle of Scandinavian mythology, enigmatic in origin, like all things evil . . .'[169] Adding him to the cast-list of his Nibelung drama, Wagner explains Loge's origin as elemental fire-spirit.[170] Perhaps he was influenced by Grimm, who defines the related Eddic giant Logi as 'the natural force of fire';[171] perhaps also by von der Hagen, for whom Loge is 'blaze ["Lohe"], light, flame—Lucifer'.[172]

The overriding influence on all matters concerning Loge, however, was undoubtedly that of Weinhold, who here writes: 'The names "Log" and "Lodr" are evidence that the god was regarded as the divinity of fire'.[173] Neither Russwurm, nor Grimm, nor even von der Hagen had persuaded Wagner to include Loge among his 1848 pantheon. Only after reading Weinhold's 'Die Sagen von Loki', which appeared in the *Zeitschrift für deutsches Alterthum* between the time of *Siegfrieds Tod* and Wagner's exit from Dresden, and embarking on new reading from the Eddas, was Wagner apparently able to conceive the demigod's role in his drama.

All in all, Weinhold's 'Loki' suggests that von der Hagen at his most untrammelled was the author's spiritual godfather. It is a work of wide-ranging erudition which in the course of

---

[168] 'Sie stellen ein untergegangnes oder untergehendes geschlecht dar, dem mit der kraft auch die unschuld und weisheit des alterthums, mehr eine objektive und anerschafne, als selbsterworbne vernunft beiwohnt.' (pp. 495–6.)

[169] 'das tiefste Räthsel der skandinavischen Mythologie, unerklärlich in seinem Ursprung, wie alles böse . . .' (*Nordische Sagen*, p. 256).

[170] See e.g. *Schr.*, vi. 84.

[171] 'die naturkraft des feuers' (*Mythologie*, p. 221).

[172] 'Lohe, Licht, Flamme—Lucifer' (*Die Nibelungen*, p. 47).

[173] 'Die namen Log und Lodr bekunden wie unser gott als die gottheit des feuers gedacht wurde' ('Loki', p. 8).

its ninety-five pages covers such topics as the vestal virgins and the relationship of the Indian god Agni with cows. Nevertheless, certain important trends in Weinhold's argument stand out clearly: Loge as fire-god, his ethical position, his function in the society of the gods, and the changing perspective of his role through the centuries. Above all, Weinhold was concerned to restore the god to what he regarded as his original stature and dignity. Loge survives in the stories Snorri tells and in certain Eddic poems which, despite episodes of unaccountable evil, tend in the main towards the comic, even the salacious. But the 'unworthy creature' here revealed, 'the court jester of Asgard, who entertains the king, Lord Odin, by gossip and foolery and who sets evil on foot wherever he can',[174] is in Weinhold's view only a late, debased representation of the god. Chief mischief-maker and embroiler of the gods, giver of specious counsel, 'evil and cunning',[175] 'clever and crafty',[176] 'a sly, seductive miscreant'[177] in the opinion of other scholars: for Weinhold the harm Loge causes is well-directed, systematic, and purposeful: 'our Loki becomes the bearer of physical and ethical destruction.'[178] Loge is a god with a mission, albeit a negative one, and it is that which lends weight to the demigod's knavish activities.

Apart from his viewpoint as a scholar, Weinhold also brought more Loge material to Wagner's attention through introducing him to the Eddic *Oegisdrecka*, the poem in which Loge cites the peccadilloes of every member of the company. Often scurrilous in tone, the *Oegisdrecka* nevertheless perfectly displays the mocking, taunting dialogue that Loge adopts when out to ruffle the gods. Weinhold thought so highly of the poem as a source that he included a paraphrase, rather than a translation, in his article.[179] Since then the

---

[174] 'unwürdiges wesen'; 'hofnarr von Asgard, der den könig, herrn Odin, durch klatschereien und possen unterhält, der böses anstiftet, wo er kann' (Weinhold, 'Loki', p. 73).

[175] 'böse und hinterlistig' (Russwurm, *Nordische Sagen*, p. 30).

[176] 'witzig und verschlagen' (Ettmüller, *Vaulu-Spá*, p. 61).

[177] 'ein schlauer, verführerischer bösewicht' (Grimm, *Mythologie*, p. 221).

[178] 'unser Loki wird der träger der physischen und ethischen vernichtung.' ('Loki', p. 28).

[179] 'Loki', pp. 67–71.

Oegisdrecka had appeared in Simrock's *Edda*, giving Wagner a chance to remind himself of Loge's showmanship.

Loge finally makes his way into the *Ring* in *Das Rheingold* to mock, beguile, ensnare, and destroy his fellows, friend and foe alike. Alberich, to be sure, suggests that Loge is more dangerous to his friends:

> Den Lichtalben
> lacht jetzt Loge,
> der listige Schelm:
> bist du Falscher ihr Freund,
> wie mir Freund du einst war'st—
> haha! mich freut's!
> von ihnen fürcht' ich dann nichts.[180]

> Loge now laughs
> with the light-elves,
> the artful rogue.
> If, false one, you're their friend
> as once you were friend to me—
> ha ha! I'm glad!
> then I've nothing to fear from the gods.

But it is a false sense of security which Alberich soon has cause to rue.

Loge's is a peformance of brilliant artistry. First he teases. Freia has been pledged to the giants, on Loge's advice, in return for the building of Walhall. The day of reckoning arrives. The anxious gods and impatient giants are kept waiting for Loge, who has promised to look for a way of redeeming the pledge. When tensions have risen to a suitable pitch the tardy Loge appears. Yes, he agrees, he promised to look for a solution: he did not promise he would find one:

> Mit höchster Sorge
> d'rauf zu sinnen,
> wie es zu lösen,
> das—hab' ich gelobt:
> doch das ich fände,
> was nie sich fügt,
> was nie gelingt,
> wie liess sich das wohl geloben?[181]

---

[180] *Schr.*, v. 242.
[181] *Schr.*, v. 223–4.

> With the utmost concern
> to study the question
> of how to resolve it:
> that—was my solemn vow.
> But that I would find
> what never can fit,
> what never can prosper—
> how could I promise that?

Next, he tempts. Loge's eulogy on womanhood has his whole audience spellbound; into this pool of attention he drops the poisoned bait of the Rhinegold and stands back to observe while gods and giants bite.[182]

Then comes his tricking of Alberich, and finally the destruction of Fasolt: his 'elementary advice'[183] to Fasolt to take the ring leads immediately and directly to the fatal quarrel with Fafner.[184] Loge completely outmanœuvres the opposition in the game he is playing, which only Wotan has wit enough to begin to appreciate and which not even Wotan is wily enough to withstand. With such feeble resistance, small wonder that Loge quickly grows contemptuous of the gods:

> Fast schäm' ich mich,
> mit ihnen zu schaffen . . .[185]

> I'm almost ashamed
> to share in their schemes . . .

Arch-tempter and taunter of the gods, Loge needs one thing more to crown his triumph: his victims must be aware of their downfall. And so Wagner introduced another dimension, *Oegisdrecka*-inspired, to Loge's activities: he is what Weinhold terms 'the objectivized conscience'[186] of the gods; 'even where he appears wicked and destructive he is operating

---

[182] *Schr.*, v. 225 ff.
[183] Thus Cooke, *World*, p. 170; otherwise Cooke is very perceptive on Loge's role and functions.
[184] *Schr.*, v. 264.
[185] *Schr.*, vi. 267.
[186] 'dies objectivierte gewissen' ('Loki', p. 71).

against a backcloth of justice . . .'[187] Each time he succeeds
in corrupting the gods he holds up a moral mirror so that
they can view their own deformity. Having persuaded the
gods to bargain away Freia he is eloquent in praise of 'the
joy and worth of woman';[188] having aroused desire for the
Rhinegold in all hearts he is tireless in pressing the Rhinemaiden's claim for its return.[189]

Even before the curtain rises on *Das Rheingold* the gods are
hopelessly enmeshed in Loge's web of perplexity and guilt.
The more they struggle, the faster they find themselves, until
eventually in the second act of *Die Walküre* Wotan ceases
resistance and accepts defeat:

> Fahre denn hin,
> herrische Pracht,
> göttlichen Prunkes
> prahlende Schmach!
> Zusammen breche,
> was ich gebaut!
> Auf geb' ich mein Werk;
> eines nur will ich noch:
> das Ende ——
> das Ende![190]

> Depart then for ever,
> lordly pomp,
> shameful parade
> of godly splendour!
> Let what I built
> break and crumble:
> I herewith renounce my work.
> I want only one thing now:
> the end ——
> the end!

Their sense of guilt lames the gods, Weinhold argues, and the
real function of *Oegisdrecka* is to show that their deficiencies
make the gods unviable: 'It reveals in the clearest possible
manner that the deities had to perish . . . through their

---

[187] 'Selbst wo er schlecht und verderblich erscheint hat er zum hintergrunde die gerechtigkeit . . .' (Weinhold, 'Loki', p. 67).

[188] 'Weibes Wonne und Wert' (*Schr.*, v. 225).

[189] *Schr.*, v. 226, 229, 233, 260, 268.

[190] *Schr.*, vi. 42.

awareness of guilt the gods have become totally impotent.'[191]
Guilt and inadequacy were for Wagner's contemporaries the
cause of the doom of the gods, and not fate.[192]

Having failed the fire-test of Loge's seductions the gods are
now ripe for their own extinction: 'After Loki has destroyed
the ethical support of the gods' edifice he turns against its
physical stability too.'[193] Already at the end of *Das Rheingold*
the 'Götterdämmerung' is in preparation. As the gods mount
to Walhall with faintly ridiculous pomp and ceremony Loge
observes:

> Ihrem Ende eilen sie zu,
> die so stark im Bestehen sich wähnen.[194]

> They are rushing to their ruin,
> who believe they can stand so steadfast.

With contempt for their delusions and pretensions he muses:

> zur leckenden Lohe
> mich wieder zu wandeln,
> spür ich lockende Lust.
> Sie aufzuzehren,
> die einst mich gezähmt,
> statt mit den blinden
> blöd zu vergeh'n—
> und wären's göttlichste Götter—
> nicht dumm dünkte mich das![195]

> I sense a desire
> to resolve myself
> into flickering flames once more,
> destroying those
> who once restrained me,
> instead of perishing
> with the blind, stupidly—
> and were they the godliest gods—
> that doesn't seem so silly!

[191] 'Es zeigt aufs klarste dass der götterkreis untergehen muste . . . die götter sind durch ihr schuldbewusstsein völlig ohnmächtig' ('Loki', p. 71).
[192] See also Simrock, *Edda*, p. 338, n. to *Völuspá*, ver. 31, and Ettmüller, *Vaulu-Spá*, p. xlv.
[193] 'Nachdem Loki den ethischen halt des göttergebäudes vernichtet hat, kehrt er sich auch gegen den physischen.' (Weinhold, 'Loki', p. 61.)
[194] *Schr.*, v. 267.
[195] Ibid.

And so in the new ending for *Siegfrieds Tod* Brünnhilde, about to light Siegfried's funeral pyre, sends Wotan's waiting ravens past the rock where Loge still glimmers to tell him the due moment has come for him to consume Walhall and all its occupants. 'The spirit of destruction and of atoning, cleansing fire'[196] comes to purge the gods of the guilt he has enticed them into. The waters of the Rhine overflow to much the same effect, for as Jacob Grimm remarks, 'All the elements are purifying, healing, atoning'.[197]

The destruction of the firmament through fire and water at the end of the era of the gods was a standing feature of Eddic mythology, expounded most fully in the *Völuspá* and in the *Snorra Edda*. Before the final flood and conflagration, however, the Eddas tell of the Last Battle, in which the gods and heroes are ranged against the massed forces of the underworld and associates and in which all perish. If Wagner omitted the battle, he did not altogether forget it; in the *Mythus* the Last Battle was definitely on the agenda, though not immediately, for Siegfried tells the Rhinemaidens: 'I know three wiser women [the norns] than you; they know where the gods will once fight in fear and dismay . . .'[198] In Act II of *Die Walküre* Wotan is still anticipating trouble from Alberich:

> Durch Alberichs Heer
> droht uns das Ende . . .[199]
>
> Alberich's army
> threatens the end for us . . .

His garnering of heroes in Walhall is in preparation for the final conflict:

> Das stark zum Streit
> uns fände der Feind,
> hiess ich euch Helden mir schaffen . . .[200]
>
> That the foe should find us

---

[196] 'der geist der vernichtung und des sühnenden reinigenden feuers' (Weinhold, 'Loki', p. 55).

[197] 'Alle elemente sind reinigend, heilend, sühnend' (*Mythologie*, p. 549).

[198] 'Drei weisere Frauen, als ihr kenne ich, die wissen wo die Götter einst in banger Sorge streiten werden . . .' (*Skizzen*, p. 31).

[199] *Schr.*, vi. 39.

[200] Ibid.

strong for the struggle
I had you gather me heroes . . .

If in the final *Ring* version the gods instead sit glumly
awaiting the end[201] and go down without a murmur, we
must look for the reason in the composer's perennial flir-
tation with the 'release from existence' theme,[202] which
makes their further existence philosophically undesirable; or
else in the impotence of the gods brought on by their guilt,
which makes resistance impossible; or indeed in the
exigencies of the theatre, which ruled that a red glow on the
horizon was both quicker and easier to muster than the
battlefield of Vígríd.

The right moment for Loge to commence the physical
destruction of the gods comes as Brünnhilde sets Siegfried's
funeral pyre alight:

> *Brünnhilde*
> Denn der Götter Ende
> dämmert nun auf:
> so—werf' ich den Brand
> in Walhalls prangende Burg.[203]

> *Brünnhilde*
> For the doom of the gods
> is dawning now:
> thus—do I kindle the fire
> in Walhall's shining fort.

Out of the fire and the flood the Rhinemaidens will take back
their gold from Brünnhilde's hand. Their debt discharged,
nothing now holds the gods and they are free to depart in
peace.

Wagner has sought to underpin the sequence of events with
ethical and philosophical causality, perhaps not altogether
successfully: almost immediately his good friend Röckel was
asking 'why, since the Rhinegold has been returned to the

---

[201] See Waltraute's account, *Schr.*, vi. 201–3.
[202] Recurrent in Wagner's dramas from the Dutchman through to Amfortas and
Kundry.
[203] *Schr.*, vi. 254.

Rhine, do the gods still have to perish?'[204] Wagner did not really have an answer, and to Röckel's query could only reply: 'the conviction that their annihilation is necessary . . . grows out of our innermost feelings.'[205]

The real connection between Siegfried's death and the twilight of the gods is not logical but poetic, and had already been established by the Romantic scholars. In the *Nibelungenlied*, Siegfried's death is followed by the 'Nibelungen Not'—the 'Nibelungs' peril'—in which Gunther and Hagen and all their clan are annihilated and their enemies likewise. After Baldur's death in the Eddas comes the twilight of the gods, in which they and all their foes perish. It was easy enough for a scholar such as von der Hagen to see parallels between the 'Nibelungen Not' and the twilight of the gods; and since for von der Hagen Siegfried was Baldur, the 'Götterdämmerung' belonged as much in the aftermath of Siegfried's death as it did in the wake of Baldur's.

It is in the coupling of Siegfried's death and the 'Götterdämmerung' that Hermann Schneider suggests we have the strongest evidence of Wagner's debt to Romantic scholarship, and most of all to von der Hagen.[206] His judgement is difficult to quarrel with. In so far as it is possible to speak of a major theme in a work of the nature of von der Hagen's *Die Nibelungen*, the major theme of its central section is the Oneness of the *Nibelungenlied* events and those of Eddic mythology. At page 37 he introduces his theme: 'But it can be demonstrated that in our tradition too Siegfried's life and death, the lament, and the Nibelungs' peril . . . are none other than the life and death of Baldur the Good and the annihilation of all the gods in the "Götterdämmerung" . . .'[207] For another hundred pages he develops the argument

---

[204] The question is set out thus by Wagner in his letter replying to Röckel: 'warum nun, da das Rheingold dem Rhein zurückgegeben ist, die Götter doch noch untergehen?' (25 Jan. 1854, *Dokumente*, p. 93.)

[205] 'Aus unserm innersten Gefühle erwächst uns . . . die Notwendigkeit dieses Unterganges.' (Ibid.)

[206] 'Altertum', p. 119.

[207] 'Aber es lässt sich darthun dass auch bei uns Siegfrieds Leben und Tod, die Klage, und der Nibelungen Noth . . . nichts anders ist, als das Leben und der Tod Baldurs des Guten, der Untergang aller Götter in der Götterdämmerung . . .'

with facts and figures and parallels and similarities. Even supposing Wagner had not read von der Hagen's *Die Nibelungen* he still knew its argument, though not from the enthusiast but from a detractor. Karl Lachmann flays von der Hagen's theory in his own 'Kritik':

> How irresponsible it was, then, arbitrarily to foist upon the saga the opinion—which could be supported only by impoverished word-plays and ill-founded notions—that the Nibelungs' peril was an illustration of the end of the world and that the whole thing [the *Nibelungenlied*] presents the origin, life, fall, death and resurrection myth of the first men or gods![208]

A secondary target of Lachmann's criticism, as so often, may have been Mone, whose *Einleitung*, as Golther points out,[209] also contains material which offends against Lachmann's views.

What then? Did Wagner ever feel with von der Hagen that 'this twilight of the gods is simultaneously the dawn of a great and everlasting day, when gods and men will no longer be different'?[210] Surely at some stage he agreed with Jacob Grimm that 'just like the Great Flood, the destruction of the world by fire is not intended to destroy for ever but to purify, and to bring in its wake a new and better world order'.[211] In a drama as complex as the *Ring*, to find one ending which would accommodate all its separate elements eventually proved an impossible task. Wagner wrote three: restoration of the patriarchy, establishing the new order of love, and escape from existence. In the end he abandoned all attempts to verbalize the conclusion of the *Ring*, leaving music as its final expression.

---

[208] 'Wie leichtsinnig war es also, der sage willkürlich die ansicht aufzudringen, die sich nur mit armseligen wortspielen und leeren einfällen unterstützen liess, der Nibelunge noth sei ein bild des weltendes, und das ganze stelle den mythus vom ursprung leben sünde tod und wiedergeburt der ersten menschen oder götter dar!' (p. 348).

[209] *Grundlagen*, p. 95.

[210] 'dieser Abenddämmerung der Götter zugleich die Morgendämmerung eines grossen ewigen Tages [ist], wo Götter und Menschen nicht mehr verschieden sind' (*Die Nibelungen*, p. 140).

[211] 'Gleich der sintflut soll auch der weltbrand nicht für immer zerstören, sondern reinigen und eine neue, bessere weltordnung nach sich ziehen' (*Mythologie*, p. 776).

9
Conclusion

THE preceding pages have been spent tracing specific instances of where the interest of Wagner's contemporaries in the Nibelungs influenced his own concept of the material and, in turn, the *Ring*. By now we have seen something of the broad range of responses the Nibelungs evoked during the successive revivals of popularity in the first half of the nineteenth century, which conserved, explored, and recreated the cherished heritage in art, literature, and scholarship; and we have become aware of some of the themes in the great and occasionally acrimonious debates that raged between scholars of divergent persuasions. We observed in Part I that Wagner sampled widely of the Nibelung wares on offer; in Part II it has become apparent that not all were ultimately of equal value to the author of the *Ring*.

In the course of the previous chapters the names of certain individuals who exercised tangible influence on Wagner's Nibelung drama stand out from the host of scholars active on the contemporary scene. Among the older generation there were Jacob Grimm, whose *Deutsche Mythologie* was Wagner's handbook for all things mythological and whose joint ventures with his brother Wilhem, in particular the *Märchen*, deeply benefited Wagner's drama; von der Hagen, who single-handed supplied half of Wagner's Nibelung source material—the *Völsunga saga* and *Thidreks saga*, *Das Lied vom Hürnen Seyfrid*, and other heroic poems—and whose arch-Romantic study, *Die Nibelungen, ihre Bedeutung für die Gegenwart und für immer*, we prefer to believe Wagner knew; Fouqué, whose *Sigurd* was initially Wagner's main access to the *Völsunga saga* and whose solutions to the problems of dramatizing the Nibelung saga made a manifestly indelible impression on the *Ring* author; and finally, by the Master's own reckoning, Franz Joseph Mone, the exact importance of

whose *Untersuchungen* may by now be somewhat less
opaque.

Of the second wave of Nibelung writers, Simrock, 'the
saga researcher and reviver I so much admire',[1] was the one
outstanding influence on Wagner. Simrock's *Amelungenlied*
offered arguably the most far-reaching reworking of the
Nibelung saga, and access to much other renovated source
material besides; the appearance of his *Edda*, the first full
translation, in the spring of 1851 may well, we have argued,
have been the catalyst which sparked off the conversion of
the work of 1848 into the *Ring*.

Surrounding this constellation of leading influences on
Wagner's Nibelung drama is a galaxy of lesser ones: Jacob
Grimm's brother and co-editor Wilhelm, author of the *Deuts-
che Heldensage*; Karl Lachmann, whose pungent 'Kritik' was
highly useful to Wagner and who supplied one of his
Nibelungenlied editions; Frauer and Weinhold, whose
influence penetrated deeper into the philosophy of the *Ring*
than the limited subjects of their monographs would suggest.
Sometimes influence was relatively limited in scope:
Raupach's drama, Russwurm's *Nordische Sagen*, Uhland's
Siegfried poem. In the case of Müller's *Altdeutsche Religion*
and Göttling's *Nibelungen und Gibelinen* we still cannot say
with perfect confidence that Wagner knew them. There
were, finally, all those other editors, translators, and illustrators of primary sources who contributed to Wagner's know-
ledge: for the *Nibelungenlied* Vollmer, perhaps Hinsberg,
Pfizer and his two illustrators, Schnorr von Carolsfeld and
Neureuther; Mohnike, rather than Wachter, as translator of
the *Heimsringla*; and those whose Edda selections served
Wagner until Simrock's translation appeared: the *Snorra Edda*
translations of Rühs and Majer, the 'Götterlieder' samples of
Majer, Studach, and Legis and the Grimm brothers'
'Heldenlieder', and the *Vaulu-Spá* and Nibelung cycle of
Wagner's Zurich friend Ettmüller.

These were the contemporaries whose interest in the
Nibelungs influenced Wagner's drama. The ways in which

---

[1] 'der so sehr von mir geschätzte Sagen-Forscher und Erneuerer' (*ML*, p. 314).

their influence worked itself out in the *Ring* can be summed up as follows. In the first place, the contemporary movement provided Wagner with his primary source material in a form he could understand, be it in original language or translation or, where these failed, paraphrase or reworking. Although not perhaps an independent contribution from his contemporaries, its importance cannot be overstressed, for without knowledge of the Nibelung saga and other related source material the *Ring* could not have been written.

Secondly, Wagner's awareness of saga events was supplemented by the inventions he found in the work of other creative artists—or indeed scholars, since Göttling's version of Siegfried's dying oration must come into this category (see Chapter 6). Sometimes the imaginative efforts of contemporaries can still be seen on the surface of Wagner's drama, like gems embedded in its outer fabric: the dialogue patterns of the *Siegfried* smithy scene, the Rhinemaidens coming for their gold in the concluding dumb show of the *Ring*, or the business of Siegfried's horse in Act 1 of *Götterdämmerung*. This is not to suggest deliberate or conscious plagiarism on Wagner's part, merely to show that such features encountered in his search were assimilated into his concept of the Nibelungs and were each slotted into place at the appropriate point of his drama.

Still at a fairly superficial level, snippets of information, accurate or otherwise, culled from the scholars produced results in Wagner's opera ranging from the 'Minnetrinken' ceremony to the life-style of the valkyries. At the same time ideas and interpretations from the various schools of Nibelung thought worked their way deep into the inner structure of the *Ring*: the unity of the heroic and the mythological worlds and the indivisibility of fairy-tale from both; Siegfried as Baldur; the many identities of Wotan; solar myth. While the philosophy of the *Ring* draws largely on Wagner's specialist reading outside the range of the Nibelung scholars, it nevertheless shows signs of having been affected by the reflections of Frauer and Weinhold on the themes of guilt and expiation and on the tormented, self-destructive world of the ancient Teuton and his gods.

Invention, information, ideas, and interpretations: all these contributions from his contemporaries influenced what material went into Wagner's Nibelung drama. They also showed him how to use it. Sometimes Wagner took quite specific guidance from the work of predecessors, as when Fouqué's scene structure found a new home in *Siegfried* or Wilhelm Grimm's selective Wälsung history inspired Wagner's terse account in the *Mythus*. But more generally, too, Wagner inherited from the Nibelung scholarship of his day ways of working his material. Wagner's genius in combining incidents and characters meaningfully, in forging links and associations where the sources failed to provide them, has been recognized and rightly admired,[2] and can be seen at work in Brünnhilde, in the Nibelungs and Alberich, in the Wälsung history. It was no doubt helped to develop by the example of Romantic scholarship, ever ready to admit identity, subordinating all to the concept of unity, adventurous in the search for Oneness.

A scholar of the age wrote that history and poetry were One,[3] and certainly the Romantics knew no rigid dividing line between research and inspiration, between scholar and poet. The creativity of Romantic scholarship, whereby the wish was often father to the thought, lives on in Wagner's *Wibelungen* essay and is one of the major generative forces at work in the *Ring*.

The final legacy of contemporary Nibelung interest can be found in Wagner's evaluation of the saga's significance. The peculiar political and psychological circumstances of German national life in the first half of the nineteenth century had retrieved the *Nibelungenlied* from the cloisters of academe and had set it in the vanguard of popular and revolutionary movements. The poem whose destiny might otherwise have been no more than to attract the aesthetic judgement and excite the philological curiosity of a few scholars became a national symbol. The Nibelung saga of love and power,

---

[2] e.g. by Golther, *Grundlagen*, p. 54; Schneider, 'Altertum', p. 114.

[3] Göttling, *Nibelungen und Gibelinen*, p. 12: 'Denn wer den Geist der Poesie vom Geist der Geschicht scheidet, hat das Seyn und Wesen dieser nie erkannt. Beide sind Eins im Wirken des Volkes . . .' ('For whoever divides the spirit of poetry from the spirit of history has failed to recognize its essence and nature. Both are One as they affect the people . . .')

greed, treachery, grief, and destruction was judged by Görres and Grün a suitable vehicle for social criticism and by von der Hagen as 'the original ancestral saga of the human race itself'.[4]

So it was that the Nibelung saga was Wagner's natural choice for his most important drama, largest in scale and richest in complexity; and that through its medium he sought to express in the *Ring* ardently held beliefs concerning love and ownership, freedom and patriarchy, and his profoundest intuitive insights into 'the very nature of the world'.[5]

---

[4] 'die Ur- und Stamm-Sage des Menschengeschlechts selber' (*Die Nibelungen*, p. 66).

[5] 'das Wesen der Welt selbst' (letter to August Röckel, 23 Aug. 1856, *Dokumente*, p. 118).

APPENDIX A

Wagner's *Ring* Text: Chronological Table

1848	4 Oct.	*Der Nibelungen-Mythus* completed
	20 Oct.	Prose plan for *Siegfrieds Tod* completed
	End of Oct.	Sketch for Norn Prologue
	12–18 Nov.	*Siegfrieds Tod*, 1st version
1849	Early	*Siegfrieds Tod*, 2nd version (*Schr.*)
	Early[a]	*Die Wibelungen*
1851	10 May	1st sketch for *Der junge Siegfried*
	24 May – 1 June	Prose plan for *Der junge Siegfried*
	3–24 June	*Der junge Siegfried*
	Early Nov.	1st sketch for *Das Rheingold*
	Mid-Nov.	1st sketch for *Die Walküre*, Acts I and II
1852	Before 23 Mar.	Further sketches for *Das Rheingold* and *Die Walküre*
	23–31 Mar.	Prose plan for *Das Rheingold*
	17–26 May	Prose plan for *Die Walküre*
	1 June – 1 July	*Die Walküre*
	15 Sept – 3 Nov.	*Das Rheingold*
	Mid-Oct – 15 Dec.	Alterations to *Der junge Siegfried* and *Siegfrieds Tod*
1853	Before 11 Feb.	*Ring* text printed privately
1856	June[b]	1st appearance of titles *Siegfried* and *Götterdämmerung*
1863		*Ring* text published, including variants from 1853 edition

All information from Strobel, except:

[a] Deathridge, Geck, and Voss, *Wager Werk-Verzeichnis*, p. 329.
[b] *Dokumente*, p. 114.

APPENDIX B

Wagner's Nibelung Studies

A. DOCUMENTED READING

1. Before the Nibelungen-Mythus *of 1848*

Works in Wagner's personal library at Dresden:
  Edda: Rühs, *Snorra Edda*
  —— von der Hagen (1812; Old Norse)
  —— Majer, *Snorra Edda* and 'Götterlieder'
  —— Ettmüller, *Vaulu-Spá*
  —— Grimm, 'Heldenlieder'
  Frauer, *Die Walkyrien*
  J. Grimm, *Mythologie* (1844)
  —— *Grammatik*
  J. and W. Grimm, *Deutsche Sagen*
  J. and W. Grimm, *Märchen*
  W. Grimm, *Heldensage*
  Snorri Sturluson, *Heimskringla*, trans. Wachter
  —— trans. Mohnike
  Heldenbuch: von der Hagen (1811; 1820–5)
  —— Simrock (1843–9, vols. i–v)
  Hürnen Seyfrid (published in *Heldenbuch*): von der Hagen (1811, 1820–5)
  —— Simrock (1843–9, vol. iii)
  Lachmann, *Zu den Nibelungen*: 'Kritik'
  Mone, *Untersuchungen*
  Nibelungenlied: Leipzig, 1840
  —— Pfizer
  —— Simrock (1843–9, vol. ii)
  —— Vollmer
  Raumer, *Geschichte der Hohenstaufen*
  Russwurm, *Nordische Sagen*
  Sachs, *Der hörnen Seufried*
  Simrock, *Das Amelungenlied*, vols. i, ii (*Heldenbuch*, vols. iv, v)
  Tieck, 'Der gestiefelte Kater'
  Uhland, 'Siegfrieds Schwert'
Not in Wagner's personal library:
  J. Grimm, *Mythologie* (1835)
  Thidreks saga (postulated reading date: pre-*Mythus*)

Loans from the Royal Library at Dresden:

17 Jan. – 18 Feb. 1844	von der Hagen, *Die Nibelungen*
11 Jan. – 1 Feb. 1845	Hinsberg, *Nibelungenlied*
1 Feb. – 1 Mar. 1845	Lachmann, *Nibelungenlied*
19 Apr. – 7 Aug. 1847	J. Grimm, *Deutsche Grammatik*
10 June – 13 Sept. 1848	{ J. and W. Grimm, *Edda*
	{ Wachter, *Heimskringla*
21 Aug. – 2 Oct. 1848	{ Legis, *Edda*
	{ Studach, *Edda*
	{ Grimm, *Deutsche Rechtsalterthümer*

2. *Between the* Mythus *of 1848 and Wagner's removal from Dresden in May 1849*

Works in Wagner's personal library at Dresden:
J. Grimm, *Geschichte der deutschen Sprache*
Simrock, *Heldenbuch*, vol. vi (*Das Amelungenlied*, vol. iii)
Weinhold, 'Loki'

Loans from the Royal Library at Dresden:

21 Oct. 1848 – 20 June 1849	{ J. Grimm, *Deutsche Rechtsalter-thümer*
	{ J. and W. Grimm, *Altdeutsche Wälder*
21 Oct. 1848 – 1 Jan. 1849	{ Ettmüller, *Edda* (1837)
	{ von der Hagen, *Völsunga saga*
10 Feb. – 19 June 1849	{ Göttling, *Geschichtliches im Nibelungenliede*
	{ Rückert, *Oberon von Mons*

B. WORKS WITHOUT DOCUMENTARY PROOF OF READING

Highly likely to have influenced Wagner:
Fouqué, *Sigurd*
Raupach, *Der Nibelungen-Hort*
Simrock, *Edda*

Probable influences on Wagner:
Göttling, *Nibelungen und Gibelinen*
Müller, *Altdeutsche Religion*

Possible influences:
Görres, *Der Hürnen Siegfried*
Mone, *Einleitung*
Vischer, 'Vorschlag zu einer Oper'

Unlikely to have influenced Wagner:
  Dumas, 'Lyderic'
  du Méril, *Histoire*
  Müller, *Versuch*
  Volksbuch: *Der gehörnte Siegfried*

APPENDIX C
Wagner's Edda Reading

	By 1848		10 June – 13 Sept. 1848	21 Aug. – 2 Oct. 1848		21 Oct. – 29 Jan. 1849	Spring 1851
	Majer	Ettmüller	Grimm	Studach	Legis	Ettmüller	Simrock
'Götterlieder'							
Völuspá	X	X		X	X		X
Grímnismál	X			X	X		X
Vafthrudhnismál				X	X		X
Skírnisför	X				X		X
Hrafnagaldr Odhins							X
Vegtamskvidha	X						X
Harbardhsliódh				X	X		X
Hymiskvidha	X			X	X		X
Oegisdrecka					X		X
Thrymskvidha	X			X			X
Alvíssmál				X			X
Fiölsvinnsmál							X
Hávamál				X			X
Grógaldr							X
Rígsmál							X
Hyndluliódh							X

Also von der Hagen's Old Norse edition of the 'Heldenlieder' and the *Snorra Edda* translations of Rühs, Majer, and Simrock.

BIBLIOGRAPHY

BENVENGA, N., *Kingdom on the Rhine* (Harwich, 1983).

BODMER, J. J., ed., *Chriemhilden Rache und die Klage* (Zurich, 1757).

COLOMB, H., 'Commentaire', in Marcel Herwegh, *Au soir des dieux* (Paris, 1933), 183–217.

COOKE, D., *I saw the world end* (London, 1979).

DAHLHAUS, C., 'Über den Schluss der Götterdämmerung', *Richard Wagner, Werk und Wirkung* (Regensburg, 1971), 97–115.

DEATHRIDGE, J., GECK, M., and VOSS, E., *Wagner Werk-Verzeichnis* (Mainz, London, New York, and Tokyo, 1986).

DREWS, A., *Der Ideengehalt von Richard Wagners dramatischen Dichtungen in Zusammenhang mit seinem Leben und seiner Weltanschauung* (Leipzig, 1931).

DUMAS, A., 'Les Aventures merveilleuses du Prince Lyderic', *Musée des familles* (Sept.–Nov. 1841).

EDDA: see under

    Ettmüller, L. (1830; 1837).

    Gräter, F. D. (1798; 1816).

    Grimm, J. L. and W. C. (1815).

    Hagen, F. H. von der (1812; 1814).

    Legis, G. T.

    Majer, F.

    Rühs, C. F.

    Simrock, K. J. (1851).

    Studach, J. L.

ELLIS, W. ASHTON, *Life of Richard Wagner*, adapted from C. F. Glasenapp, 'Das Leben Richard Wagners', 6 vols. (London, 1900–8).

—— 'Die verschiedenen Fassungen von "Siegfrieds Tod"', *Die Musik*, 3 (1904), 239–51, 315–31.

ETTMÜLLER, L., trans. *Vaulu-Spá* (Leipzig, 1830).

—— *Die Lieder der Edda von den Nibelungen* (Zurich, 1837).

FOUQUÉ, F. BARON DE LA MOTTE, 'Der gehörnte Siegfried in der Schmiede', *Europa*, 2 (1803) pp. 82–7.

—— *Der Held des Nordens* (Berlin, 1810).

—— *Sigurd der Schlangentödter, Ausgewählte Werke*, vol. i (Halle, 1841).

FRAUER, L., *Die Walkyrien der skandinavisch-germanischen Götter- und Heldensage* (Weimar, 1846).

GOLTHER, W., *Die sagengeschichtlichen Grundlagen der Ringdichtung Richard Wagners* (Berlin, 1902).

GÖRRES, G., *Der hürnen Siegfried und sein Kampf mit dem Drachen* (Schaffhausen, 1843).

GÖTTLING, C. W., *Über das Geschichtliche im Nibelungenliede* (Rudolstadt, 1814).

—— *Nibelungen und Gibelinen* (Rudolstadt, 1816).

GRÄSSE, J. G. T., trans., *Gesta Romanorum*, 2 vols. (Dresden and Leipzig, 1842).

—— *Die Sage vom Ritter Tanhäuser, aus dem Munde des Volks erzählt* (Dresden and Leipzig, 1846).

GRÄTER, F. D., ed. and trans., *Nordische Blumen* (Leipzig, 1798).

—— trans., *Hrafna Galldur Odins, Alvis-Mál: Idunna und Hermode*, 4 (1816), Nos. 34, 35, 36, 39, 44.

GRIMM, J. L. C., *Deutsche Rechtsalterthümer* (Göttingen, 1828).

—— *Deutsche Mythologie*, 2 vols. (Göttingen, 1835); 2nd edn., 2 vols. (Göttingen, 1844).

—— *Deutsche Grammatik*, 3rd. edn., pt. 1 (Göttingen, 1840).

—— *Geschichte der deutschen Sprache*, 2 vols. (Leipzig, 1848).

—— and GRIMM, W. C., eds., *Altdeutsche Wälder*, 3 vols. (Kassel and Frankfurt-on-Main, 1813–16).

—— eds. and trans., *Die Lieder der alten Edda* (Berlin, 1815).

—— eds. and trans., *Deutsche Sagen*, 2 vols. (Berlin, 1816–18).

—— eds., *Kinder- und Haus Märchen*, 2nd edn., 3 vols. (Berlin, 1819–22).

GRIMM, W. C., ed., *Die deutsche Heldensage* (Göttingen, 1829).

GRÜN, ANASTASIUS, [pseud. of Anton Alexander Graf von Auersperg], *Gedichte* (Leipzig, 1843).

—— *Nibelungen im Frack* (Leipzig, 1843).

HAGEN, F. H. VON DER, ed., *Der Nibelungen Lied nebst Glossar* (Berlin, 1807).

—— ed., *Das Nibelungenlied in der Ursprache* (Berlin, 1810); 3rd edn. (Breslau, 1820).

—— ed., *Der Helden Buch*, vol. i (Berlin, 1811).

—— ed., *Lieder der älteren oder Sämundischen Edda* (Berlin, 1812).

—— trans., *Die Edda Lieder von den Nibelungen* (Breslau, 1814).

—— trans., *Nordische Heldenromane*, 5 vols. (Breslau, 1814–28): vols. i–iii, *Wilkina- und Niflunga-Saga, oder Dietrich von Bern und die Nibelungen*; vol. iv, *Völsunga saga*; vol. v, *Ragnar-Lodbroks-Saga, Norna-Gests-Saga*.

—— *Die Nibelungen, ihre Bedeutung für die Gegenwart und für immer* (Breslau, 1819).

HAGEN, F. H. VON DER, ed., *Der Nibelungen Noth*, 3rd edn. (Breslau, 1820).

— and A. Primisser, eds., *Der Helden Buch in der Ursprache* (Deutsche Gedichte des Mittelalters, ed. F. H. von der Hagen and J. G. G. Büsching, vol. ii, pts. 1 and 2; Berlin, 1820, 1825).

HELDENBUCH: see under
Hagen, F. H. von der (1811; 1820–5).
Simrock, K. J. (1843–9).

HERMANN, F. R., *Die Nibelungen* (Leipzig, 1819).

HERMANNSSON, HALLDÓR, *Bibliography of the Eddas* (Islendica, vol. xiii; New York, 1920).

HINSBERG, J. VON, trans., *Das Lied der Nibelungen*, 4th edn. (Munich, 1838).

HÜRNEN SEYFRID, DAS LIED VOM: see under
Hagen, F. H. von der (1811; 1820–5).
Simrock, K. J. (1843–9, vol. iii).

KOCH, E., *Richard Wagner's Bühnenfestspiel Der Ring des Nibelungen in seinem Verhältnis zur alten Sage wie zur modernen Nibelungendichtung betrachtet* (Leipzig, 1876).

LACHMANN, K., ed., *Der Nibelunge Not mit der Klage* (Berlin, 1826); 2nd. edn. (Berlin, 1841).

— *Zu den Nibelungen und zur Klage* (Berlin, 1836): pp. 333–49, 'Kritik der Sage von den Nibelungen'.

LEBEDE, H., *Richard Wagners Musikdramen. Quellen, Entstehung, Aufbau*, vol. ii (Dresden, n.d.).

LEGIS, G. T., trans., *Edda* (Fundgruben des alten Nordens, vol. ii/1; Leipzig, 1829).

LICHTENBERGER, H., *Richard Wagner, poète et penseur*, 3rd. edn. (Paris, 1902).

MAJER, F., trans., *Mythologische Dichtungen und Lieder der Skandinavier* (Leipzig, 1818).

MEINCK, E., *Die sagenwissenschaftlichen Grundlagen der Nibelungendichtung Richard Wagners* (Berlin, 1892).

MÉRIL, E. DU, *Histoire de la poésie scandinave. Prolégomènes* (Paris, 1839).

MOHNIKE, G., trans., *Heimskringla: Sagen der Könige Norwegens von Snorre Sturlason*, vol. i (Stralsund, 1837).

MONE, F. J., *Einleitung in das Nibelungen-Lied zum Schul- und Selbstgebrauch bearbeitet* (Heidelberg, 1818).

— *Untersuchungen zur Geschichte der teutschen Heldensage* (Quedlinburg, 1836).

MÜLLER, J. W., *Chriemhilds Rache* (Heidelberg, 1822).

MÜLLER, W., *Versuch einer mythologischen erklärung der Nibelungen-sage* (Berlin, 1841).
—— *Geschichte und System der altdeutschen Religion* (Göttingen, 1844).
MYLLER, C. H., ed., *Der Nibelungen Liet* (Berlin, 1782).
NEWMAN, E., *The Life of Richard Wagner*, 4 vols. (paperback edn., Cambridge, 1976).
NIBELUNGENLIED
   Daz ist der Nibelunge Liet (Leipzig, 1840).
   see also under
   Bodmer, J. J.
   Hagen, F. H. von der (1807; 1810, 1820; 1820).
   Hinsberg, J. von.
   Lachmann, K. (1826, 1841).
   Myller, C. H.
   Pfizer, G.
   Simrock, K. J. (1843–9, vol. ii).
   Vollmer, A. J.
OTTO, L., 'Die Nibelungen', *Neue Zeitschrift für Musik*, 23 (1845), Nos. 13, 33, 43, 44, 46.
PFIZER, G., *Der Nibelungen Noth* (Stuttgart and Tübingen, 1843).
PFORDTEN, H. VON DER, *Richard Wagners Bühnenwerke in Handlung und Dichtung nach ihren Grundlagen in Sage und Geschichte*, 7th edn. (Berlin, 1920).
PRODHOMME, J. G., 'Une source française de l'Anneau du Nibelung', *Mercure de France*, NS, 279 (1937), 15 Oct., 439 ff.
RAUMER, F. L. G. VON, *Geschichte der Hohenstaufen und ihrer Zeit*, 2nd edn., 6 vols. (Leipzig, 1840–1).
RAUPACH, E., *Der Nibelungen-Hort* (Hamburg, 1834).
RÜCKERT, E., *Oberon von Mons und die Pipine von Nivella* (Leipzig, 1836).
RÜHS, C. F., trans., *Die Edda* (Berlin, 1812).
RUSSWURM, C., *Nordische Sagen der deutschen Jugend erzählt* (Leipzig, 1842).
SACHS, H., *Hans Sachs ernstliche Trauerspiele . . . und Possen*, ed. je J. G. Büsching, 3 vols. (Nuremberg, 1816–24): vol. ii, *Der hörnen Seufried*.
SAXO GRAMMATICUS, *Saxonis Grammatici Historia Danica*, ed. P. E. Müller, 2 vols. (Copenhagen, 1839).
—— *The First Nine Books of the Danish History of Saxo Grammaticus*, trans. O. Elton (London, 1894).
SCHNEIDER, H., 'Richard Wagner und das germanische Altertum', *Kleinere Schriften* (Berlin, 1962), 107–24.

SIMROCK, K. J., *Wieland der Schmied* (Bonn, 1835).

—— trans., *Das Heldenbuch*, 6 vols. (Stuttgart and Tubingen, 1843–9): vol. i, *Gudrun*; vol. ii, *Das Nibelungenlied*, 3rd edn.; vol. iii, *Das kleine Heldenbuch*; vols. iv–vi, *Das Amelungenlied*.

—— ed., *Die deutschen Volksbücher*, 13 vols. (Frankfurt-on-Main, 1845–67).

—— trans., *Die Edda, die ältere und jüngere* (Stuttgart and Tübingen, 1851).

SNORRI STURLUSON, HEIMSKRINGLA: see under
Monike, G.
Wachter, F.

SPENCER, S., 'Zieh hin! Ich kann dich nicht halten', *Wagner*, NS, 2 (1981), 98–120.

STROBEL, O., ed., *Richard Wagner: Skizzen und Entwürfe zur Ringdichtung* (Munich, 1930).

—— 'Zur Entstehungsgeschichte der Götterdämmerung', *Die Musik*, 25 (1933), 336 ff.

STUDACH, J. L., trans., *Sämund's Edda des Weisen* (Nuremberg, 1829).

THIDREKS SAGA: see under
Hagen, F. von der (1814–28, vols. i–iii).

THORP, M., *The Study of the Nibelungenlied* (Oxford Studies in Modern Languages and Literature; Oxford, 1940).

TIECK, L., *Ludwig Tieck's Schriften*, 28 vols. (Berlin, 1821–54): vol. v, 'Der gestiefelte Kater'.

UHLAND, L., *Gedichte* (Stuttgart and Tübingen, 1842): p. 402, 'Siegfrieds Schwert'.

VISCHER, F. T., 'Vorschlag zu einer Oper', *Kritische Gänge*, vol. ii (Tübingen, 1844), 399 ff.

VOLLMER, A. J., ed., *Der Nibelunge Nôt und diu Klage* (Dichtungen des deutschen Mittelalters, vol. i; Leipzig, 1843).

VÖLSUNGA SAGA: see under
Hagen, F. H. von der (1814–28, vol. iv).

WACHTER, F., trans., *Snorri Sturluson's Weltkreis*, 2 vols. (Leipzig, 1835).

WAGNER, R., *Der Ring des Nibelungen* (Zurich, 1853).

—— *Sämtliche Schriften und Dichtungen*, 6th edn., (Volksausgabe), 12 vols. (Leipzig, n.d.).

—— *Siegfried, Götterdämmerung*, ed. and trans. W. Mann (Friends of Covent Garden; London, 1964).

—— *Dokumente zur Entstehungsgeschichte des Bühnenfestspiels 'Der Ring des Nibelungen'*, ed. W. Breig and H. Fladt (Richard

Wagner: *Sämtliche Werke*, ed. C. Dahlhaus, vol. xxix/1; Mainz, 1976).

—— *Mein Leben*, ed. M. Gregor-Dellin (Munich, 1976).

—— *The Ring*, trans. A. Porter (Folkestone, 1976).

WEINHOLD, K., 'Die Sagen von Loki', *Zeitschrift für deutsches Alterthum*, 7 (1849), 1–94.

WESTERNHAGEN, C. VON, *Richard Wagners Dresdener Bibliothek 1842–49* (Wiesbaden, 1966).

—— *Wagner* (Zurich, 1968).

WESTON, J. L., *The Legends of the Wagner Dramas* (London, 1896).

WURM, C., *Die Nibelungen* (Erlangen, 1839).

Zeitschrift für deutsches Alterthum, ed. M. Haupt, vols. 1–7 (Leipzig, 1841–9).

INDEX

Page numbers in parentheses denote that the subject of the entry is alluded to, but not named as such.